Management of Defense Acquisition Projects

Management of Defense Acquisition Projects

Edited by
Rene G. Rendon and Keith F. Snider
Naval Postgraduate School
Monterey, California

Ned Allen, Editor-in-Chief
Lockheed Martin Corporation
Palmdale, California

Published by
American Institute of Aeronautics and Astronautics, Inc.
1801 Alexander Bell Drive, Reston, VA 20191-4344

American Institute of Aeronautics and Astronautics, Inc., Reston, Virginia

1 2 3 4 5

Library of Congress Cataloging-in-Publication Data

Rendon, Rene G. (Rene Garza), 1960–
Management of defense acquisition projects / Rene G. Rendon and Keith F. Snider, Ned Allen.
p. cm.
ISBN 978-1-56347-950-2 (alk. paper)
1. United States. Dept. of Defense--procurement. 2. United States--Armed Forces--Weapons systems--Purchasing. I. Snider, Keith L. II. Allen, Ned. III. Title.

UF503.R45 2008
355.6'2120973--dc22 2008023648

Cover design by Virginia Kozlowski

FOREWORD

The largest, financially strongest, and most vibrant innovation factory in the world is the Defense Acquisition, Technology, and Logistics (AT&L) sector—originally a U.S. creation, and now a worldwide phenomenon. It extends all the way from technology readiness levels (TRLs) supporting fundamental research at national, university, and industrial labs working at TRL 1 to fielding mature and proven high-tech systems at TRL 9. Hundreds of thousands of scientists, engineers, and other professionals are employed in the field, with an astonishing diversity of backgrounds and skills, and at an annual cost greater than the economies of some countries. That so large and complex a socioeconomic system is organized at all is truly astonishing—but it is so!—if not always perfect. The sector harbors dynamic and vital acquisition processes, hosts many of the world's most creative people and institutions, and orchestrates the most ambitious technology development projects ever envisioned: the Joint Strike Fighter, the International Space Station, the Orion Crew Exploration Vehicle, and others.

Management of Defense Acquisition Projects is a coordinated effort by the graduate faculty on acquisition at the U.S. Naval Postgraduate School in Monterey, California. Under the leadership of Rene G. Rendon and Keith F. Snider, this book has been created to serve the entire AT&L community. It is an introductory text, but it also has enough enduring value—in spite of the mercurial regulatory environment that brings so much dynamism to the AT&L community—to earn a spot on the desktop of the sector's seasoned practitioners. We are gratified to have this anchor volume join the other AT&L studies planned for and currently hosted by the Library of Flight.

The Library of Flight is part of the growing portfolio of information services from the American Institute of Aeronautics and Astronautics. It extends the Institute's publications with the best from a growing variety of topics in aerospace from aviation policy, to case studies, to studies of aerospace law, management, and beyond. The series seeks to document the crucial role of aerospace in enabling, facilitating, and accelerating global commerce, communication, and defense, and so, this reference volume on AT&L occupies a very significant position on the Library of Flight shelf. As new aerospace programs grow and change around the world, we plan for the Library to host a wide array of international authors, expressing their own points of view on aerospace visions, events, and issues. As the demands on the world's aerospace systems grow to support new capabilities such as unmanned vehicles, international relief, agricultural management, environmental monitoring, and others, the series will seek to document landmark events, emerging trends, and new views.

Ned Allen
Editor-in-Chief
Library of Flight

TABLE OF CONTENTS

PREFACE

For some years, it has been obvious to us, as teachers of defense acquisition management, that the absence of standard textbooks in the field constitutes a serious shortcoming in the educational resources available to the acquisition workforce. A precious few classics—Gansler's *Affording Defense* and Fox's *Arming America*, for example—teach students about acquisition, but few if any texts explicitly aim to equip students to serve in acquisition jobs. We intend for this book to accomplish that purpose.

In this textbook, we do not portray acquisition as a separate discipline with its own body of knowledge. Rather, we believe that acquisition is a unique field of study and practice that makes use of concepts and tools from multiple reference disciplines. Consequently, this book contains a wide variety of both theoretical and practical material on 1) the unique context of defense acquisition, 2) proven tools and techniques, 3) best practices, and 4) management concepts and their application in the acquisition domain.

We intend this principally as a survey text that covers a broad range of relevant acquisition topics, and we hope it will serve as a valuable reference for a broad range of acquisition professionals in both the public and private sectors. In addition to basic knowledge and tools that will benefit the beginning acquisition student, we also included advanced concepts to engage and challenge experienced scholars and practitioners. In essence, we wanted to present material that addresses not only "how to do" acquisition, but also how to think about doing it better.

One of our main objectives in this project was to produce a text with enduring relevance and utility. Anyone familiar with acquisition knows that its policies and procedures are notoriously volatile. Some past textbooks, although worthy when initially released, became immediately dated and largely useless once policies changed. Thus, we sought to write this book using fundamental and time-tested materials—materials that, we hope, will remain applicable regardless of any future policy shifts.

We recognize, however, that teachers of acquisition courses will want their students to be knowledgable of the most up-to-date policies and procedures. We believe that an appropriate way to resolve this tension will be to use this book as a core foundational textbook that should be supplemented with readings covering the most current policies. Acquisition teachers will also no doubt want to supplement this text with other readings, such as current reports from the Government Accountability Office (GAO) and the Department of Defense Inspector General (DoDIG), as well as articles from professional and trade associations.

We believe that the strongest feature of this book is the quality of our co-authors. All of the chapter authors are not only knowledgable veterans of the "real world" of acquisition management, but they are also students of acquisition who actively pursue research and consulting projects in their respective areas of expertise. Their chapters thus reflect their insights gained both from professional experience as well as from scholarly pursuits. Clearly, this textbook would not have been possible without their participation, and we gratefully acknowledge their contributions.

Our thanks go to Rodger Williams and Ned Allen of AIAA, both of whom provided support and advice throughout the project. We also acknowledge the support of James B. Greene (Rear Admiral, U.S. Navy, retired) and Karey L. Shaffer of the Acquisition Research Program at the Naval Postgraduate School. Most important, we thank our wives, Juanita Rendon and Kate Snider, for their patience and loving support throughout this project.

Finally, the order in which our names appear does not indicate the relative importance of our contributions. Considering ourselves equal partners in this project, we flipped a coin to determine the order.

Rene G. Rendon
Keith F. Snider
June 2008

Chapter 1

Project-Management Concepts

Rene G. Rendon* and Keith F. Snider†
Naval Postgraduate School, Monterey, California

Learning Objectives

- Understand the definition of a project
- Describe the characteristics of a project
- Understand the triple constraint and its impact on the project manager
- Differentiate between projects and routine operations
- Differentiate between project management and other disciplines
- Discuss the project life cycle and its importance in managing projects
- Understand the interdisciplinary nature of project management
- Identify the three decision support systems that affect acquisition outcomes

1.1 *Introduction*

This chapter establishes the foundation for this book by introducing several specific project-management concepts. Project management is the core disciplines underlying defense acquisition management. Therefore, it is essential to understand the specialized definitions, features, and tools of project management that distinguish it from other types of operations and make it so useful in acquisition management. In discussing these project-management concepts, numerous references will be made to *A Guide to the Project Management Body of Knowledge* (*PMBOK Guide*), published by the Project Management Institute [1], the premier professional association for project management.

---

*Graduate School of Business and Public Policy.
†Associate Professor, Graduate School of Business and Public Policy.

1.2 DEFINITION OF A PROJECT

The *PMBOK Guide* defines a project as a "temporary endeavor undertaken to create a unique product, service, or result" ([1], p. 368). The operative terms in this definition relate to the temporary and unique characteristics of projects—that is, *projects* are temporary, in that they have a beginning and an end, and they are unique, meaning that each project has characteristics that are different from the others. More about other project characteristics will be presented in the next section.

Projects are differentiated from programs in that *programs* refer to "groups of related projects that are managed in a coordinated way to obtain certain benefits and control" ([1], p. 368). Thus, program managers might manage a program, such as a weapons system program, which consists of a number of different projects, each managed by a project manager. (In defense acquisition usage, distinctions between the titles "program manager" and "project manager" are not rigorous. For example, the U.S. Marine Corps has a program manager for the Expeditionary Fighting Vehicle, whereas the U.S. Army has a project manager for the Joint Lightweight Howitzer. The U.S. Air Force uses the title system program director to refer to the program/project manager position. Thus, in this text, we will simply use the acronym PM to avoid confusion.) The term *project management* is defined as the "application of knowledge, skills, tools and techniques to project activities to meet project requirements" ([1], p. 368). Therefore, *program management* is "the centralized, coordinated management of a group of projects to achieve the program's strategic objectives and benefits" ([1], p. 368).

1.3 CHARACTERISTICS OF A PROJECT

The preceding section already identified two characteristics of projects: they are temporary and unique. Other characteristics of projects include purpose, life cycle, interdependencies, and conflict. Each will be discussed next:

Temporary: Projects are temporary in nature, as opposed to being ongoing efforts. The temporary characteristic of projects should not lead one to believe that projects are short in duration, or noncomplex, or inconsequential. *Temporary* basically means that projects have a finite life span, with a definite beginning and an end.

Unique: Projects are unique in that the project outcome, or deliverable, is different from other similar supplies or services. Every project, with its specific cost, schedule, performance objectives, and related risk environment, is individually customized. If it were not for this unique characteristic, projects would be considered routine operations.

Purpose: All projects have a purpose, typically described in terms of well-defined end results or outcomes. The PM and project team are charged with

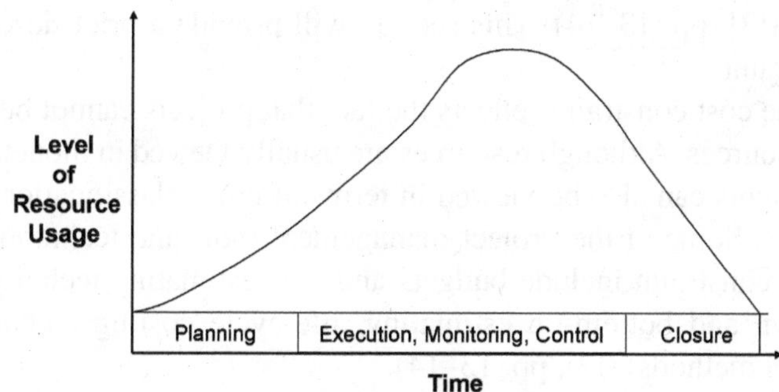

Fig. 1.1 Sample generic project life cycle (adapted from [4], p. 13).

the task of achieving those results or outcomes. Thus, project management involves a purpose-driven work environment and culture.

Life cycle: It was stated earlier that all projects are temporary, with a definite beginning and a definite end. Related to this finite nature of projects is the life-cycle characteristic of projects. The project life cycle refers to the various stages of activity or phases through which the project progresses on its way from beginning to completion. These phases might be separated by managerial decision points called control gates (also known as milestone decisions). Additionally, many projects reflect a "slow-rapid-slow" progress pattern. That is, the project starts off slowly, then gathers momentum as the work gets accomplished, and finally slows down as the work tapers off near completion ([2], pp. 14, 15). Figure 1.1 illustrates this aspect of the project life cycle, as well as generic phases of a project. These project phases will be discussed later in this chapter.

Interdependencies: All projects are composed of various tasks that need to be accomplished in order for a PM to meet the project objectives. These tasks are interdependent of each other and must be accomplished in a certain sequence. As these tasks are performed in an interdependent manner, they compete for the same resources (such as time, material, money, and personnel), which typically results in conflict ([2], p. 10).

Conflict: As the project tasks are performed in accordance with the project schedule, they compete with each other for the same limited resources and personnel. The PM's job is focused on resolving conflict from the competing areas of the project—namely, cost, schedule, and performance.

1.4 THE TRIPLE CONSTRAINT

PMs are responsible for completing the project within *budget*, on *time*, and according to the *specifications*. These three areas of the project reflect the basic PM's goals: achieving the *cost*, *schedule*, and *performance* objectives of the project. Cost, schedule, and performance are considered the important parameters of project management and are often referred to as *the triple constraint* ([3], pp. 13–14). This section will provide a brief description of each constraint.

Cost: The cost constraint reflects the fact that projects cannot be managed without resources. Although resources are usually viewed in monetary terms, cost constraints can also be viewed in terms of other classifications, such as person-days. Some of the project-management tools and techniques related to the cost constraint include budgets and cost-estimating techniques (such as top-down and bottom-up estimating, life-cycle costing techniques and cost-control methods) ([3], pp. 13–14).

Schedule: The schedule constraint reflects the time dimension of projects. As PMs strive to complete projects "on time," they are challenged with dealing with task-duration estimates, task dependencies, and schedule conflicts. Project-management tools and concepts dealing with these schedule constraints include Gantt charts, network diagrams, critical paths, schedule slack, and schedule crashing.

Performance: The performance constraint reflects the major purpose of the project, and that is to develop or produce a solution to satisfy a customer's need. The performance requirements of a project are critical to the success of the project itself. History is full of examples of projects completed according to performance specifications, but which were not successful because they did not satisfy the needs of the customer [3]. Some of the tools available to help PMs deal with the performance constraint include work breakdown structures (WBS), statements of work (SOW), specifications, and standards.

The triple constraint is considered the focal point of the PM's attention as he/she manages the project. A useful analogy is that of the PM skillfully juggling three balls in the air—each ball represents one of the constraints. The PM does not necessarily want to optimize any one specific constraint, but wants to manage all three constraints and conduct tradeoffs throughout the project period to achieve an optimal solution.

It is intuitively obvious that cost, schedule, and performance are all positively correlated. For example, achievement of a high level of performance requires more time and money than a lower level of performance. Of course, it is rarely possible in any project for all three constraints to grow unconstrained at the same time. Rather, PMs usually develop, in concert with the customer, minimally acceptable performance levels, ceiling costs, and required delivery dates that constitute a "trade space" (i.e., a space within which tradeoffs are possible) of acceptable outcomes. Figure 1.2 illustrates the triple-constraint concept and the space in which the PM can manage tradeoffs, integrating cost, schedule, and performance objectives to successfully complete the project.

Projects that do not allow the PM flexibility in applying tradeoffs to the triple constraints would be considered overconstrained projects and not suitable for the application of project-management concepts and tools.

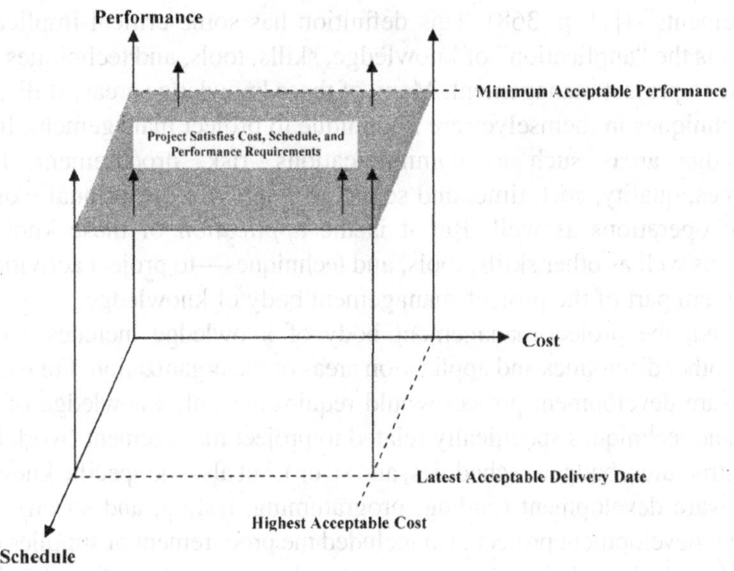

Fig. 1.2 Triple constraint and a "trade space" of acceptable outcomes.

1.5 *PROJECTS COMPARED TO ROUTINE OPERATIONS*

As stated earlier in this chapter, projects are temporary and unique endeavors with a definite beginning and a definite end. These major characteristics distinguish projects from nonprojects. Nonproject work, called routine operations or operational work, differs from projects in that it is ongoing, repetitive, and routine ([1], p. 6). Although routine operations might share some similar characteristics with projects, they are fundamentally different because of their objectives. Routine operations are focused on continuously sustaining the mission of the organization, whereas projects are focused on attaining the project objectives and then terminating. An example of a project differing from routine operations would be the case of a U.S. Navy aircraft carrier. The combined acquisition, development, and production of the aircraft carrier was a project—with a definite beginning with blueprints and drawings and a definite end when it was launched at sea. However, the maintenance of the ship while it is at sea and at port would be considered routine operations, focused on sustaining the mission of the fleet. The recent emergence of the project-management discipline and profession is a result of the realization that projects cannot be adequately managed by the same management principles and techniques used for routine operations ([2], p. 11).

1.6 *PROJECT MANAGEMENT COMPARED TO OTHER DISCIPLINES*

Recall the definition of project management as the "application of knowledge, skills, tools and techniques to project activities to meet project requirements" ([1], p. 368). This definition has some critical implications. First, it is the "application" of knowledge, skills, tools, and techniques that is specific to project management. Many of these knowledge areas, skills, tools, and techniques in themselves are not unique to project management. Indeed, knowledge areas such as communications, risk, procurement, human resources, quality, cost, time, and scope also apply to operational work and routine operations as well. But it is the *application* of these knowledge areas—as well as other skills, tools, and techniques—to project activities that make them part of the project-management body of knowledge.

Second, the project-management body of knowledge includes, to some extent, other disciplines and application areas of the organization. For example, a software development project would require not only knowledge of skills, tools, and techniques specifically related to project management (work breakdown structures, budgets, schedules, and so on), but also the specific knowledge of software development (coding, programming, testing, and so on). If this software development project also included the procurement of supplies or services, then the knowledge of procurement and contracting practices would also be required by the project team. Thus, project management is related to the particular applications involved in the specific project. Also related to project management are the general business-management knowledge areas related to the routine and ongoing operations. These general business-management areas consist of the planning, organization, directing, and controlling processes of the organization. Therefore, the project-management body of knowledge—indeed, the boundaries of the profession for that matter—encompass, to some extent, the particular application knowledge areas related to project work (engineering, software management, contracting, production, logistics, and so forth), as well as the general business-management knowledge areas, in addition to the specific project-management knowledge areas.

Finally, we should include one last comment on project-management knowledge, concepts, and skills. Project-management concepts and skills are universally applicable to different disciplines and fields of work. The concepts of work breakdown structures, budgets, and schedules transcend many industry boundaries and are "industry independent." However, PMs are not "industry independent." Effective PMs not only have to have strong competencies in project-management knowledge areas, but must also have strong technical skills in their respective application area or discipline. Thus, in order to be an effective software PM, one must also be an effective software manager ([5], pp. 15–17).

1.7 Interdisciplinary Nature of Project Management

Earlier, we introduced the three objectives (or triple constraint) of cost, schedule, and performance, which must be balanced for success in any project. These three objectives help demonstrate the inherently interdisciplinary nature of project management. That is, all but the most trivial of projects entail the need for several different areas of specialized, technical, and managerial knowledge and skills in order to meet their cost, schedule, and performance objectives.

To illustrate this point, consider three relatively simple projects: a family vacation, a home landscaping project, and a company picnic. Table 1.1 lists only a few of the possible cost, schedule, and performance aspects involved in each.

Clearly, knowledge and skills in these different aspects of cost, schedule, and performance will contribute to success in each of these projects. For example, knowledge of the range of available lodging options should increase the probability that the family will be satisfied with its accommodations throughout its vacation. For the landscaping project, knowledge of and skills in designing retaining walls, walkways, and water features will increase the likelihood that the homeowner will be pleased with the final project outcome.

The requisite skills and knowledge for successful attainment of project objectives can come from a variety of sources. The preceding simple projects might often be do-it-yourself efforts in which no outside expertise is required or obtained (e.g., the company employees themselves provide food and music for the picnic, rather than having a catered event with a hired band). Alternatively, outside experts might be brought in for some aspects of the project, such as, for example, a loan officer to arrange a home equity loan for the landscaping project, or a travel agent to arrange details of the vacation schedule.

As a project's complexity increases, so also does the breadth and depth of knowledge and skills necessary for project success. For example, a new home construction project is rarely a do-it-yourself effort; rather, it requires contributions from a wide range of experts, such as architects, general contractors, excavators, masons, carpenters, roofers, electricians, and plumbers. Participation by local government officials might also be required for permits and architectural reviews. Thus, increased project complexity leads to increased interdisciplinary complexity, often significantly raising the difficulty of attaining cost, schedule, and performance objectives.

TABLE 1.1 SKILLS AND KNOWLEDGE AREAS FOR SIMPLE PROJECTS

Project	Cost	Performance	Schedule
Family vacation	Affordability, financing	People to visit, sights to see, meals, means of transportation, accommodations	Dates of travel, duration of stay
Home landscaping	Affordability, financing	Design, materials, construction, permits	Sequence of activities, deliveries of materials
Company picnic	Affordability, financing	Activities, food, music, cleanup	Date selection, site reservation

1.8 PROJECT LIFE CYCLE

We already discussed the characteristics of a project and identified the life-cycle characteristic of projects in Sec. 1.3. The life cycle, with its phases and control gates, is an extremely valuable means for managing projects. The project life cycle provides the project team with a defined management approach for the sequential phases of a project. This defined management approach benefits the project team by providing a reliable road map for directing and monitoring the activities of the project, as well as control gates for providing a system of project monitoring and control [6]. The road map reflects the sequential phases of the project, with each phase consisting of specific processes, work activities, and tools used by the project team. The control gates provide the PM and higher-level decision makers with a mechanism for regulating and controlling the progress of the project. As the project progresses through each sequential phase, the completion of each phase should result in specific project-phase deliverables and outputs. These deliverables should be reviewed and evaluated to determine if the planned objectives of the project phase were achieved. Based on the results of each project-phase review, the project decision-making authority will determine if the project should progress into the next phase, remain in the current phase until satisfactory completion of the project-phase deliverables or outputs, or be terminated. Thus, the benefits of developing and utilizing a project life cycle include a road map for sequencing project activities, as well as control gates for regulating the progress of the project.

Project life cycles can be different and unique to each organization and for different types of projects. Depending on both the type of projects being performed by the organizations, as well as the industry in which the organization operates, project life cycles are tailored to meet the specific demands of the of the organization's environment, constraints, customers, and stakeholders. For example, the project life cycle used by a construction firm (Fig. 1.3) would be different from the project life cycle used by a major CPA firm for conducting internal control audits (Fig. 1.4). Likewise, both differ from the life cycle for defense acquisition projects to be discussed later.

1.9 LIFE CYCLE FOR DEFENSE ACQUISITION PROJECTS

In the last half-century, U.S. government policies and regulations have promoted several variations of a model life cycle for defense acquisition projects. Specific details of these life cycles have changed over the years as policy makers have attempted to reform and improve the processes for managing these projects. Yet, the basic structure of the U.S. defense acquisition project life cycle has remained remarkably unchanged since 1976. In that year, Office of Management and Budget (OMB) Circular A-109, "Major Systems Acquisition," specified four critical control gates or milestone decisions for federal acquisition projects [7]: 1) identification and definition of a specific

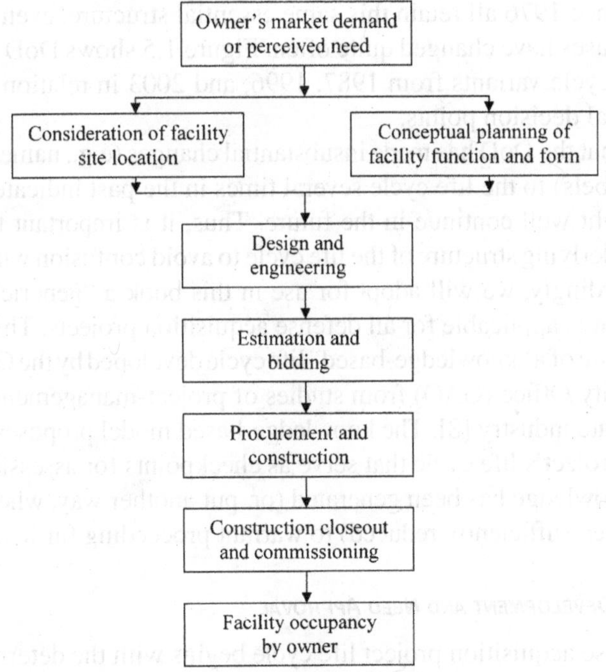

Fig. 1.3 Representative construction project life cycle.

need to be fulfilled by the acquisition, 2) selection of design concept(s) for development, 3) commitment to full-scale development and limited production, and 4) commitment to full production.

These control gates define, through the activities necessary to achieve them, a relatively stable set of phases for the defense acquisition project life cycle. The life-cycle variants that the U.S. Department of Defense (DoD) has

Determine audit requirement Establish time frame Develop audit strategy Consider industry and regulatory environment	Interview personnel Observe specific controls Inspect documents Trace transactions Select internal controls to test	Identify control objective Evaluate risk of control failure Sample transactions Assess risk of misstatement Assess fraud prevention or detection	Evaluate control function Evaluate personnel authority and qualifications Document test and evaluation results	Form opinion: unqualified, qualified, disclaimer, or adverse Prepare report
Plan the Audit	Select Internal Controls	Test and Evaluate Internal Control Design and Operating Effectiveness		Document Opinion

Time →

Fig. 1.4 Representative internal control audit project life cycle.

promoted since 1976 all retain this same essential structure, even though the names of phases have changed quite often. Figure 1.5 shows DoD acquisition project life-cycle variants from 1987, 1996, and 2003 in relation to the four A-109 critical decision points.

The fact that the DoD has made insubstantial changes (e.g., names of phases, milestone labels) to the life cycle several times in the past indicates that such changes might well continue in the future. Thus, it is important to learn the essential underlying structure of the life cycle to avoid confusion when changes occur. Accordingly, we will adopt for use in this book a "generic" life cycle (Fig. 1.6) that is applicable for all defense acquisition projects. This life cycle is an adaptation of a "knowledge-based" life cycle developed by the Government Accountability Office (GAO) from studies of project-management best practices by private industry [8]. The knowledge-based model proposes a series of points in a project's life cycle that serve as checkpoints for assessing whether sufficient knowledge has been generated (or, put another way, whether uncertainty has been sufficiently reduced) to warrant proceeding further.

1.9.1 *Need Development and Need Approval*

The defense acquisition project life cycle begins with the determination of needs and desired capabilities to be satisfied through acquisition of materiel or services (see Chapter 3). Representatives of the operational forces usually lead this effort. The need or desired capability is documented, reviewed,

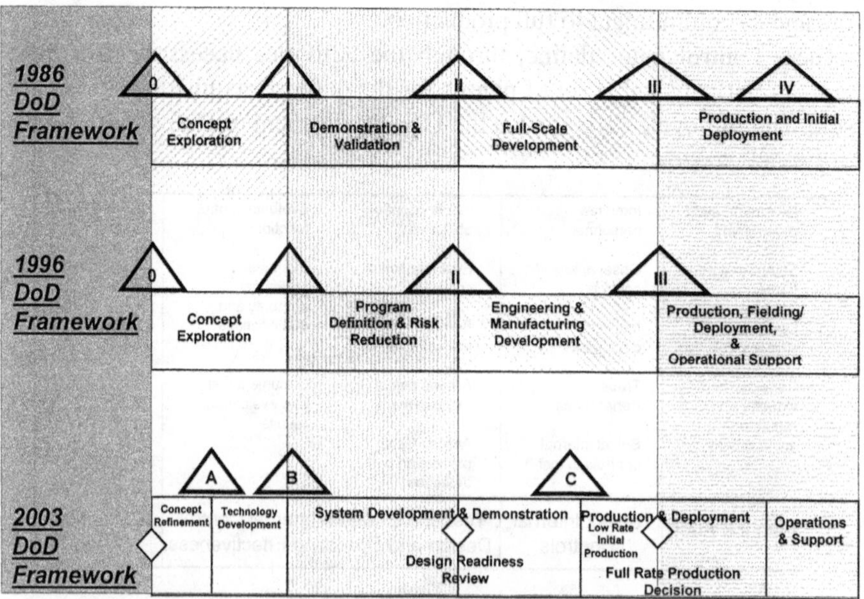

Fig. 1.5 DoD project life cycles and critical decision points.

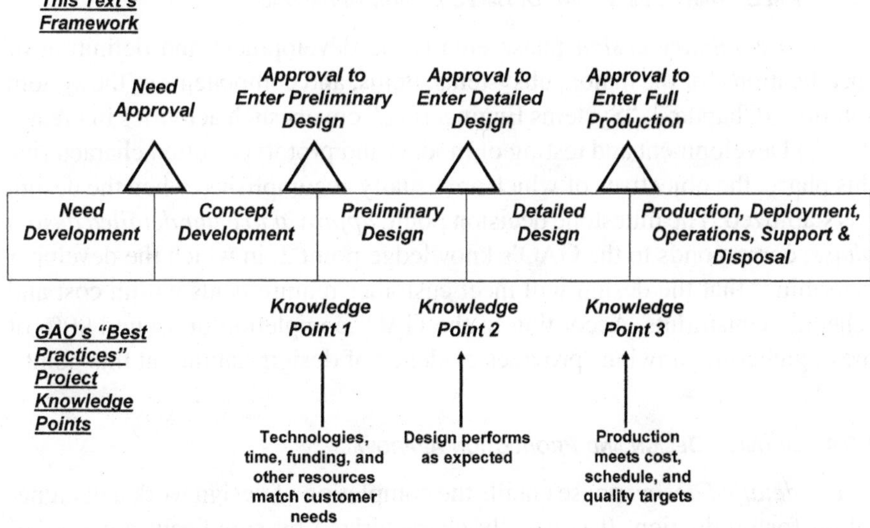

Fig. 1.6 This text's project life cycle.

validated, and approved. A favorable *need approval* milestone decision by acquisition officials begins the process of determining the most appropriate solution to satisfy that need.

1.9.2 CONCEPT DEVELOPMENT AND PRELIMINARY DESIGN APPROVAL

Concept development entails studying and refining preferred concept(s) to satisfy the approved need. This phase often entails cost-benefit and tradeoff analyses of various solutions, as well as development and assessment of technologies necessary to achieve those solutions. It culminates with the *preliminary design approval to enter* milestone decision, which corresponds to the GAO's knowledge point 1. At this milestone, decision makers review the "business case" for the preferred solution, that is, the match between the need and the resources (knowledge, time, workforce, money) necessary to acquire the preferred solution. The business case would be presented in the *acquisition strategy*, a key document that describes the program's management approach and that serves as the basis for planning and executing all aspects of the program.

Approval at this milestone decision usually represents the formal initiation of a new acquisition program in DoD, and it approves proceeding with design of the preferred solution. Another key managerial element of the program that would be approved at this milestone is the *acquisition program baseline*, which describes the program in terms of its cost, schedule, and performance parameters. As mentioned earlier, these are typically given with both threshold (minimally acceptable) and objective values in order to create space for tradeoffs in achieving acceptable outcomes.

1.9.3 Preliminary Design and Detailed Design Approval

The *preliminary design* phase entails the development and definition of specifications for the major subsystems, items, and components of the system solution. (Chapter 4, "Systems Engineering," covers such activities in greater detail.) Development and testing of models and prototypes often characterize this phase, the objectives of which are usually accomplished when the design has stabilized. The milestone decision point, *approval to enter detailed design phase*, corresponds to the GAO's knowledge point 2, in which the developer determines that the design will meet customer requirements within cost and schedule constraints. According to the GAO, completion of around 90% of the engineering drawings provides evidence of design stability at this point.

1.9.4 Detailed Design and Production Approval

The *detailed design* phase entails the completion of design work and preparation for production. It is usually characterized by significant amounts of testing on all aspects of system performance, to include reliability and other support-oriented testing. It also involves demonstrations of production processes, capabilities, and capacities. The *approval to enter full production* milestone corresponds to the GAO's knowledge point 3, at which sufficient knowledge has been achieved to ensure that the product can be produced within cost, schedule, and performance goals.

In defense acquisition projects, it is common for managers to include a *low rate initial production* (LRIP) phase for items of which a large number will be manufactured. LRIP reduces risks associated with transitioning from design to manufacturing, and it also provides production-representative items for testing.

During the *production, deployment, operations, support, and disposal* phase, the system is manufactured, deployed to the user, and supported in its operations. Testing can continue, and modifications can be undertaken, which might result in a new project to manage the modification(s). After some time, the continual process of *need development* will result in an assessment that the system is no longer adequate or necessary and will perhaps also result in a new project for a replacement system or capability. Disposal can take several forms, from scrapping items to selling them to other nations.

1.9.5 Tailoring the Project Life Cycle

The defense acquisition project life cycle is not intended as a rigid set of phases and decisions to be followed in "lock-step" fashion. Rather, it should be adapted or *tailored* to fit any project's particular requirements. This means that phases and decision points can be adjusted or eliminated if they are not needed. Consider a project to acquire new transport aircraft, for example. If, during *concept development*, the preferred concept that emerges is modifying an existing commercial airliner, then the *preliminary development* phase for development might not be necessary. [This is an example of acquiring a commercial off-the-shelf (COTS) product. COTS products are types of non-developmental items (NDI), a category of items whose designs already exist. Acquiring COTS products and NDI usually allows for tailoring the project life cycle by eliminating or shortening the design phases.] It might be possible in such a case to move directly into *detailed design* to focus on development, integration, and demonstration of the military-unique modifications to the commercial models, as well as to focus on preparations for production. Allowances for tailoring of the project cycle recognize that each defense acquisition project is—by necessity and by design—unique.

1.9.6 EVOLUTIONARY DEVELOPMENT STRATEGIES

A project following the life cycle shown in Fig. 1.6 provides the full capability required by the customer following completion of all of the phases. But what if the customer needs capability immediately or if requirements are not well known? *Evolutionary acquisition* aims to increase responsiveness to the customer by providing capability in successive increments, beginning with some limited initial increment that can be fielded quickly. In those projects where the requirement cannot be determined with finality, this strategy also helps the customer to define and refine requirements through the employment of these initial capability increments.

1.10 COMPLEXITY OF DEFENSE ACQUISITION PROJECTS

Earlier, we stated that increased project complexity leads to increased interdisciplinary complexity—often significantly raising the difficulty of attaining cost, schedule, and performance objectives. Defense acquisition projects rank among the most complex of all projects. Much of this complexity stems from the unique environment in which these projects are carried out. First, this environment has a *military* aspect. Defense acquisition projects are carried out for military purposes; thus, they have performance outcomes associated with critical "life and death" issues of national security and defense. Second, these projects often incorporate highly *advanced technologies* that are presumed to contribute to military advantages. Thus, "high tech" research and engineering skills usually play very important roles in defense acquisition projects. Third, these projects are *public* in that they are managed by public officials and financed with public funds. As a consequence, they are subject to political processes of resource allocation, review, and oversight (e.g., scrutiny of individual projects by the U.S. Congress). This unique environment thus places unique requirements on defense acquisition projects in terms of the range of skills and knowledge needed to carry out these projects successfully.

1.10.1 Important Acquisition Disciplines

Contemporary defense acquisition project management first began to be practiced in large military aerospace projects in the 1950s [9]. Entailing an unprecedented level of complexity, these projects required an exceptional integration of diverse managerial and technical skills and disciplines for success. Since that era, the most widely practiced skills and disciplines that contribute to attainment of defense acquisition project objectives include the following.

- *Systems engineering*—the principal technical discipline that ensures proper design and construction of any defense system to satisfy user or customer requirements
- *Test and evaluation (T&E)*—the systems-engineering "subprocess" for verification and validation

  It receives special attention in defense acquisition project management because of statutory requirements that specify certain procedures be followed for defense-system testing.
- *Production management*—the effective use of resources to produce the necessary amount of end units that meet quality, performance, cost, and schedule requirements

  Production management is usually very important in defense acquisition projects because of the inherent difficulty of producing large numbers of complex defense systems, which requires huge investments of money, manpower, materials, and industrial facilities.
- *Acquisition logistics*—activities that address supportability considerations throughout the acquisition life cycle

  Adequate logistical support of defense systems is essential for those systems to perform effectively in the field. In many past defense acquisition projects, however, support considerations often were not addressed early enough nor to a sufficient degree. The performance problems that resulted have led to significant emphasis on acquisition logistics.
- *Contract management*—buying goods and services (via a contract)

  Almost all defense systems and items of equipment are obtained by the government from private-sector sources. A contract is typically the vehicle that governs the transaction. Thus, contract-management knowledge and skills are of central importance in defense acquisition projects.
- *Financial management*—includes cost estimation, budgeting, cost management, accounting, and cost control

  The huge monetary investments required for most defense acquisition projects place a premium on the importance of these financial-management disciplines.

Obviously, there are other skills and knowledge areas that are also important for successful defense acquisition, and these will be addressed at appropriate places throughout the rest of this book. The disciplines just described, however, occupy such central roles that they will each have individual chapters that give them more detailed treatment.

1.10.2 THREE DECISION SUPPORT SYSTEMS FOR SUCCESSFUL ACQUISITION OUTCOMES

Acquisition is only one of the DoD's many processes and systems that enable it to accomplish its mission. Effective interaction between and among all of these is essential for DoD's success. Specifically, the two systems that must work most closely in consonance with the acquisition system in order for acquisition to be successful are as follows:

1) The *requirements system* (see Chapter 3) is operated principally by uniformed military agencies and staffs (e.g., the Joint Chiefs of Staff) that represent operational forces to determine and promote their required capabilities, many of which are provided through acquisition projects.

2) The *resource management system* (see Chapter 11) is operated by comptroller and financial-management agencies and offices. It accomplishes planning and budgeting to provide necessary funding for acquisition programs.

Acquisition managers "own" neither of these systems. Rather, they must work closely with their counterparts in each system to ensure that their projects address valid operational needs and have sufficient resources to satisfy those needs.

1.11 CONCLUSION

This chapter has introduced project management and the features that make it critical to defense acquisition management. As subsequent chapters present and explain acquisition's various interdisciplinary aspects just mentioned, they will continue to reinforce the many ways in which effective project management contributes to achieving successful acquisition outcomes.

REFERENCES

[1] Project Management Inst., *A Guide to the Project Management Body of Knowledge (PMBOK Guide)*, Newtown Square, PA, 2004.

[2] Meredith, J. R., and Mantel, S. J., *Project Management: A Managerial Approach*, 6th ed., Wiley, Hoboken, NJ, 2006, pp. 10–15.

[3] Frame, J. D., *Project Management Competence*, Jossey-Bass, San Francisco, 1999, pp. 13–14.

[4] Project Management Inst., *A Guide to the Project Management Body of Knowledge (PMBOK Guide)*, Newtown Square, PA, 2000, pp. 9–15.

[5] Verzuh, E., *The Fast Forward MBA in Project Management*, Wiley, Hoboken, NJ, 1999, pp. 15–17.

[6] Forsberg, K., Mooz, H., and Cotterman, H., *Visualizing Project Management*, 3rd ed., Wiley, Hoboken, NJ, 2005, p. 85.

[7] Office of Management and Budget, "Major Systems Acquisition," Circular A—109, Washington, D.C., 1976.
[8] Government Accountability Office, "Implementing a Knowledge-Based Acquisition Framework Could Lead to Better Investment Decisions and Project Outcomes," GAO-06-218, Washington, D.C., Dec. 2005.
[9] Baumgardner, J. S., *Project Management*, Irwin, Homewood, IL, 1963, pp. 4–8.

STUDY QUESTIONS

1.1 What characteristics of projects and of project management set them apart from other activities and operations?

1.2 Why is the triple constraint (cost, schedule, and performance) so critical in project management?

1.3 How does the project life cycle contribute to effective management?

1.4 Which disciplines, other than the ones identified in Sec. 1.10, would be important in defense project management?

1.5 Think of a major activity (e.g., sports event, church picnic) that you have recently been involved in planning and executing. To what extent were (or could) some of the project-management concepts covered in this chapter employed in this activity?

Chapter 2

DEFENSE ACQUISITION'S PUBLIC POLICY IMPRINT

KEITH F. SNIDER*
NAVAL POSTGRADUATE SCHOOL, MONTEREY, CALIFORNIA

LEARNING OBJECTIVES

- Understand the relationship between "policy" and "politics"
- Understand the importance of public policy in defense acquisition management
- Describe various makers of acquisition policy and their roles
- Describe ways in which acquisition policy originates
- Describe various ways in which acquisition policy is expressed
- Describe several areas of acquisition policy

2.1 *INTRODUCTION*

This chapter describes the importance of understanding defense acquisition policy for successful program management. Acquisition policies emerge for a variety of reasons from a variety of sources and take a variety of forms. Some policies might help acquisition officials implement their desired managerial strategies; other policies might restrict or confound their efforts. It is useful to think of a "policy highway" with a wide range of conditions: some stretches broad, smooth, and well lit; others dark and narrow; others bumpy; and some perhaps impassable. In this metaphor, the acquisition professional must understand the "policy road conditions" in order to steer programs to a successful conclusion.

The word *policy* refers to plans or courses of action that guide organizational decisions and actions, which are usually stated in some authoritative document or pronouncement. *Public policy* denotes those developed by instruments and agencies of government. Consider the following:

1) Addressing the nation in March 1983, President Reagan announced the Strategic Defense Initiative (SDI), which launched a multibillion dollar effort

---

*Associate Professor, Graduate School of Business and Public Policy.

to develop and deploy a highly effective missile defense system for the United States.

2) In the fall of 1990, Congress passed legislation known as the Defense Workforce Improvement Act (DAWIA), which directed and made provisions for the establishment of a professional acquisition workforce in the U.S. Department of Defense (DoD).

3) In May of 1995, Secretary of Defense William Perry directed the DoD to implement integrated product and process development (IPPD), a management approach intended to integrate all acquisition activities from concept through production and support, using multifunctional integrated products teams (IPTs).

Each of these policies has had enormous impacts on defense acquisition in the United States: SDI, in terms of the acquisition *programs* pursued by the DoD; DAWIA, in terms of the career development of the *people* who work in acquisition; and IPPD, in terms of the *processes* used to acquire new systems and capabilities. Such policies abound to the extent that defense acquisition can be described as having a distinct *policy imprint*. Thus, it is impossible to understand acquisition apart from its myriad policies that direct, guide, enable, constrain, and prohibit public officials and employees in their daily decisions and activities.

2.2 POLICY AND POLITICS

It would be a mistake, however, to see all of acquisition policy as forming a rational, coherent, stable, or comprehensive architecture for decisions and actions. Rather, policy is guided largely by politics; therefore, public policy necessarily has a political component. One political scientist has defined politics as "the authoritative allocation of value" [1], while Howard Lasswell's definition is given in the title of his book, *Politics—Who Gets What, When, How* [2]. Politics necessarily involves conflict and uses of power, and those in power adopt policies to accomplish some political purpose. Because political purposes often shift and because people and parties move in and out of power, it is not surprising that acquisition policies are not more rationally aligned and structured.

What purposes does defense acquisition serve? Most visibly, it provides equipment and services in support of defense missions. But because defense acquisition entails such large amounts of dollars and other resources, national leaders use it as a tool to further other-than-military goals, such as social and economic goals. A few examples of policies that reflect such diverse goals include the following:

- Preferences for domestic products in the 1933 Buy American Act (41 USC 10)
- Preferences in awarding contracts to small businesses, women-owned small businesses, disabled veterans-owned businesses, and small businesses in historically underutilized business zones

- Limitations on the DoD's allocation of depot weapon systems maintenance work between the public and private sectors (10 USC 2466)

This complex policy environment is what sets acquisition in the public sector apart from procurement in the private sector. Private firms are typically concerned about procurement's effect only on bottom-line profitability. Defense acquisition officials are likewise concerned with issues of cost effectiveness, but they are also obliged to serve a variety of policy objectives, some of which might actually work against "the bottom line" (Box 2.1).

> **BOX 2.1 BERETS AND THE BERRY AMENDMENT [3]**
>
> To protect the nation's industrial base in wartime, Congress in 1941 enacted the Berry Amendment, which gives procurement preference to domestic sources of various products. A major political controversy involving the Berry Amendment arose in 2000 after the Army Chief of Staff, General Eric Shinseki, announced that soldiers would wear black berets as standard headgear. To procure the berets, Defense Logistics Agency (DLA) waived the Berry Amendment, finding that no domestic supplier could produce berets in sufficient quantities to satisfy the Army's needs. Subsequently, contracts were awarded to firms in China, Romania, Sri Lanka, and other countries, and to one U.S. firm. The ensuing outcry from domestic firms, veterans' groups, and members of Congress led General Shinseki to direct that U.S. troops would not wear berets made in China or with Chinese content. This resulted in the disposal of over 900,000 berets, valued at $6.5 million, as well as a policy directive by the Secretary of Defense that further waivers to the Berry Amendment must be approved at his level.

Of course, not all acquisition policies are explicitly political. Many policies are concerned mainly with efficiency and effectiveness in acquisition management. For example, in the 1970s the DoD began to be concerned about the proliferation of computer software programming languages in its weapon systems, many of which were becoming obsolete and difficult to maintain. In 1975, a working group was formed to investigate the feasibility of developing and using a standard programming language in the DoD. This led to the Ada programming language and its adoption in 1987 by the DoD as its preferred language. With the advent of object-oriented programming and other software advances, however, the DoD recognized that a standard language was no longer desirable. It subsequently rescinded the preference for Ada in 1997.

2.3 ACQUISITION REFORM IMPERATIVE

Reform is acquisition policy's defining theme. Criticisms of high costs, excessively long schedules, and poor performance are perennial in acquisition programs, along with occasional findings of waste, fraud, and abuse. Throughout U.S. history, numerous reform initiatives have sought improvements, with varying degrees of success:

Hoover Commissions (1949/1955)
McNamara initiatives (1961)
Commission on Government Procurement (1972)
OMB Circular A-109 (1976)
Defense Acquisition Improvement Program (1981)
Grace Commission (1983)
Packard Commission (1986)
Goldwater–Nichols Defense Reorganization Act (1986)
Defense Management Review (1989)
Defense Acquisition Workforce Improvement Act (1990)
Section 800 Panel Report (1993)
Federal Acquisition Streamlining Act (1994)
Federal Acquisition Reform Act (1996)
Services Acquisition Reform Act (2002)
Defense Acquisition Performance Assessment (2005)

The fact that reform efforts continue today is evidence that lasting reform has been elusive.

Why is acquisition reform so problematic? The answer probably lies in the complex and dynamic nature of the policy environment just described. Consider what everyone, including political leaders, the military forces, and taxpayers, expects of defense acquisition management: 1) to provide effective and suitable systems and services for military missions; 2) to provide these in a business-like way, emphasizing cost and process efficiency; 3) to do both of the preceding in a political environment, satisfying multiple policy goals; and 4) to do all of the preceding in a world of rapid and continual change.

The inherent difficulty of the acquisition task indicates that there are continual opportunities for improvement, and so we should expect reform policies to continue. For the same reason, however, we should consider reform a continual process rather than an achievable end-state.

2.4 POLICY MAKERS

The broad authorities and responsibilities for acquisition policy making in the United States are found in the Constitution, Articles I and II of which concern the principal actors: Congress and the Executive Branch.

2.4.1 CONGRESS

Article I clearly states the Congress's powers regarding acquisition: "To raise and support armies [...] To provide and maintain a navy [...] To make rules for the government and regulation of the land and naval forces [... and] To exercise exclusive legislation in all cases whatsoever [...] for the erection of forts, magazines, arsenals, dock-yards, and other needful buildings" [4].

These powers have roots in the Continental Congress's activities and experiences during the Revolutionary War. Congress took such interest in acquisition matters that it formed several committees to procure arms and other war materials. Following the Constitution's ratification in 1789, Congress established the position of Purveyor of Public Supplies, who administered federal contracts under the Treasury Secretary Alexander Hamilton, along with an elaborate oversight and reporting system to monitor and control his activities.

Given the Constitution's language, it is unsurprising that Congress has historically been and remains very active in acquisition policy making. Congress has enacted literally hundreds of acquisition-related bills on a broad range of topics in the last half-century. An example of sweeping legislative policy is the Goldwater-Nichols Department of Defense Reorganization Act of 1986, which included major changes to the DoD's acquisition decision-making structures, affecting almost every facet of defense acquisition. By contrast, an example of more narrowly focused legislation is a provision of the 1985 Defense Authorization Act that established requirements for warranties in defense systems. (Congress later repealed this requirement in 1997 in the face of arguments that such warranties were expensive, unnecessary, and probably unenforceable). Three major pieces of acquisition legislation in the last two decades are as follows:

- *Federal Acquisition Streamlining Act* (*FASA*) of 1994, which replaced the existing small purchase threshold of $25,000 with a simplified acquisition threshold of $100,000, and created a "micropurchase" category for purchases below $2,500
- *Federal Acquisition Reform Act* (*FARA*) of 1996, which was intended to enable the federal acquisition system to emulate successful buying practices used in the commercial marketplace, by reforming larger cost and information technology (IT) acquisitions
- *Services Acquisition Reform Act* (*SARA*) of 2003, which, recognizing the growing importance services acquisition, allows for those under $25 million value to be treated as streamlined acquisitions

Congress has several resources to assist in its legislative policy-making function. Personal and committee staff members, many of whom have professional acquisition experience, provide advice and draft legislation. The Government Accountability Office (GAO, formerly General Accounting Office) serves as an investigative arm of Congress by performing accounting functions, conducting inquiries, and preparing reports on matters of congressional interest. The GAO has, in recent years, published several dozen reports dealing with defense acquisition, many of which have provided influential criticisms of the DoD's management of acquisition programs. Finally, the Congressional Research Service, part of the Library of Congress, supports members of Congress and their staffs with scholarly research on topics of interest.

2.4.2 Executive Branch

Article II of the Constitution vests acquisition policy making with the President, albeit indirectly:

- As the holder of "executive power"
- As "commander in chief of the Army and Navy"
- In the requirement to "recommend to [Congress's] consideration such measures as he shall judge necessary and expedient"

Of course, the President actually makes relatively little acquisition policy. The vast majority of Executive Branch policy is made by dozens of lower-level leaders and managers. These include political appointees—secretaries, undersecretaries, assistant secretaries, directors, administrators, and so on—as well as hundreds of career civil-service executives.

The Office of Federal Procurement Policy (OFPP), part of the Office of Management and Budget (OMB), was established by Congress in 1974 to provide guidance for federal procurement. It promulgates acquisition policies and maintains the *Federal Acquisition Regulation* (*FAR*) [5], which provides the basic rules for acquisition throughout the federal government. The OFPP administrator is appointed by the President and confirmed by the Senate.

Other important policy makers in the Executive Branch are military leaders. An important acquisition role of senior uniformed officers, who represent the ultimate users of acquired systems and services, is to make policy regarding processes by which needed capabilities and system requirements are identified and validated.

2.4.3 Policy "Influencers"

A variety of other groups influence acquisition policy making. Industry, as one of the main participants in defense acquisition, lobbies government officials to make policies that favor its interests. Firms contribute financially to various political campaigns and causes. They also band together in associations [for example, National Defense Industrial Association (NDIA)] to increase their clout. Additionally, not-for-profit "think tanks" (e.g., RAND) perform studies and analyses, some of which might influence acquisition policy. Their activities are typically funded by grants and corporate contributions. Finally, the media often find acquisition-related stories newsworthy, particularly those that highlight apparently egregious problem areas (Box 2.2). Such attention by the media, whether warranted or not, focuses the public's demands for reform policies in those problem areas.

BOX 2.2 HORROR STORIES?

In 1985, the conservative *National Review* gave its spin on the defense procurement "horror stories" then pervading the media:

"Who has not heard of the $435 claw hammer, the $640 toilet-seat cover, the $659 ashtray, and the $3,046 coffee maker? Forests of newsprint are being felled and thousands of kilowatts of precious energy consumed to vent congressional and media outrage over runaway defense costs. Support for defense spending erodes with each new tale of horror.

With good reason, the horror stories never amount to much more than the name of a commonplace product and a dollar figure. If the stories were told in full, there would not be much horror left. Consider the following epilogues:

- It is true that the Defense Department paid $435 for a claw hammer. The error was discovered by a Navy employee, and the contractor refunded the price. Last year the department bought more than eighty thousand hammers at between $6 and $8 each.
- The $640 toilet-seat cover was not a toilet-seat cover at all, but a heavy molded plastic cover for the entire toilet system of the P-3 aircraft. The toilet seats themselves cost only $9.37 each. The contractor, moreover, refunded 85 per cent of the cost of the plastic covers when the department stopped the purchase.
- Finally there is the much-publicized case of the $3,046 coffee maker. This was for an airplane, the C-5, that can carry up to 365 troops. Delta and TWA buy similar coffee makers for $3,107 each. Scratch one horror story.

Given that the Defense Department each year signs some 13 million contracts with more than 300,000 contractors, it is not unreasonable to suppose that an occasional horror story will turn up despite the best efforts and intentions. The irony is that virtually every case of serious fraud and abuse which the media have gloried in of late has been uncovered by the Defense Department's own Office of the Inspector General. [Secretary of Defense] Weinberger was pilloried for cleaning up an act he inherited.

> The real horror story may be what Congress itself has done to the Defense Department. In its zeal to achieve the utopia of an error-free, fraud-free, waste-free defense budget, it has added thousands of laws, procedures, and hurdles to the process of budgeting; it has burdened the Pentagon with literally hundreds of thousands of requests for information; it has mandated layer upon layer of new bureaucracy, overseers, watchdogs, and safeguards for the system. Weapons programs must now run a gauntlet of paperwork so burdensome and devious as to add far more to their cost than is ever saved by the safeguards. If Congress continues its crusade, we can expect that . . . not a single case of waste, fraud, or abuse will be reported anywhere in the department and not a single weapon will be procured. Total control resulting in total immobility" [6].

2.5 POLICY ORIGINS

Policies arise in several different ways. First, a policy maker might have a specific agenda for solving problems or addressing issues. For example, in 1993, then President Clinton announced the National Performance Review (NPR), to be led by Vice President Gore, "to create a government that works better and costs less" [7]. Over the next several years, Gore championed many of NPR's "reinventing government" efforts, which led to several significant acquisition policy initiatives (e.g., FASA and FARA, as already mentioned) in the mid- to late-1990s.

Second, some "trigger event" might heighten awareness of some need that can be satisfied through a policy. For example, the increase in coalition casualties in Iraq from improvised explosive devices (IEDs) led Congress to pass legislation authorizing the Secretary of Defense to implement "rapid acquisition" processes. Such processes bypass many of the usual acquisition procedures, thus allowing for faster deployment of promising capabilities to the operational forces.

Finally, policy is created for reasons that are more mundane. The internal DoD acquisition policy directive, known as 5000.1, is usually reissued as presidential administrations change. Although the new documents supposedly represent each new incoming administration's direction and priorities, the broad policies of the 5000.1 reflect generally uncontroversial goals, such as promoting flexibility and innovation; thus, these have remained remarkably consistent over time.

2.6 POLICY EXPRESSIONS

The discussion to this point has suggested several ways in which policy is expressed: laws, regulations, executive directives, and agency operating documents.

2.6.1 *LAW*

In law, acquisition policy is enacted through the usual legislative process—complete with hearings, testimonies, debates, amendments, and votes. This process often entails conflict and compromise, reflecting political tensions among members of Congress and between the Congress and the Administration over control of acquisition. For example, in the National Defense Authorization Act for fiscal year (FY) 1982, Congress included the so-called Nunn–McCurdy Amendment, which, among other provisions, called for the termination of any acquisition program whose projected unit cost-estimates exceeded original estimates by more than 25%, unless the Secretary of Defense provided various certifications to Congress that the program should continue. In passing this legislation, Congress asserted its oversight role, while restricting the Secretary's authority and managerial flexibility.

As indicated in the "media horror stories" Box 2.2, Congress is often criticized for "micromanaging" the DoD by enacting very specific oversight requirements in its acquisition legislation (Box 2.3). However, given acquisition's seemingly perennial problems, it is perhaps not so surprising that members of Congress find such measures necessary considering their views of their legitimate constitutional authorities and responsibilities for defense acquisition.

Occasionally, however, acquisition legislation passes with a remarkable degree of consensus and cooperation between the Administration and both houses of Congress. For example, in 1996, both FARA and the Information Technology Management Reform Act (later known together as the Clinger-Cohen Act) enjoyed almost unanimous support because they would allow the federal procurement system to use many successful private-sector practices.

2.6.2 *REGULATIONS*

Regulations governing acquisition include the *FAR*, under the purview of OFPP, and the *DoD FAR Supplement* (*DFARS*). Together, these regulations enumerate policies, procedures, and rules for acquiring goods and services from the private sector. They also implement provisions of some statutes, such as the Competition in Contracting Act (CICA, Public Law 98-369). Both the *FAR* and *DFARS* are considered *administrative law* governed by the provisions of the Administrative Procedures Act, passed in 1946. This act

BOX 2.3 CONGRESSIONAL "MICROMANAGEMENT"?

The following is an extract from the National Defense Authorization Act for FY 1996 on the Army's Crusader fire support system:

"Not later than August 1, 1996, the Secretary of the Army shall submit to the congressional defense committees a report containing documentation of the progress being made in meeting the objectives set forth in the baseline description for the Crusader program [...] The report shall specifically address the progress being made toward meeting the following objectives:

(1) Establishment of breech and ignition design criteria for rate of fire for the cannon of the Crusader.

(2) Selection of a satisfactory ignition concept for the next prototype of the cannon.

(3) Selection, on the basis of modeling and simulation, of design concepts to prevent chamber piston reversals, and validation of the selected concepts by gun and mock chamber firings.

(4) Achievement of an understanding of the chemistry and physics of propellant burn resulting from the firing of liquid propellant into any target zone, and achievement, on the basis of modeling and simulation, of an ignition process that is predictable.

(5) Completion of an analysis of the management of heat dissipation for the full range of performance requirements for the cannon, completion of concept designs supported by that analysis, and proposal of such concept designs for engineering.

(6) Development, for integration into the next prototype of the cannon, of engineering designs to control pressure oscillations in the chamber of the cannon during firing.

(7) Completion of an assessment of the sensitivity of liquid propellant to contamination by various materials to which it may be exposed throughout the handling and operation of the cannon, and documentation of predictable reactions of contaminated or sensitized liquid propellant."

At the time, the Army planned to field Crusader beginning in 2005 ([8], p. 22).

ensures constitutional safeguards by requiring regulating agencies, including the DoD, to keep the public informed of their procedures and rules and to provide for public participation in their rule-making processes. The *FAR* and the *DFARS* thus allow for a common understanding of requirements and expectations in contracting between government and the private sector.

Accordingly, contracting officials are not at liberty to ignore or to violate the *FAR* and the *DFARS*.

OMB Circular A-76, "Performance of Commercial Activities," is another important Executive Branch policy document that has major implications for defense acquisition. The federal government has long relied on the private sector for commercial services. A-76 requires that commercial activities should be subject to competition to ensure value for cost (Box 2.4). Accordingly, agencies must, among other requirements, do the following:

- Identify all activities performed by government personnel as either commercial or "inherently governmental," that is, so intimately tied to the public interest as to require performance by public employees
- Use competition, under the *FAR*, to determine if government personnel should perform a commercial activity

Box 2.4 A-76 Cost Savings

Gansler and Lucyshyn [9] report that, on average [in A-76 competitions], the winning bid (either public or private) leads to sustained savings of more than 30% of the projected total costs, with no decrease in performance. When examined historically, it appears that average savings have increased over time because of improvements in handling these competitions. Average savings before 1994 were around 31%; savings from competitions since then have averaged around 42%.

2.6.3 *Orders and Directives*

Within the Executive Branch, leaders at all levels make decisions, set direction, and issue instructions on acquisition matters. Two examples of presidential orders, Reagan's SDI and Clinton's NPR, were cited earlier; Figure 2.1 provides an example from the Carter Administration.

The OFPP administrator, as the President's policy chief for the Executive Branch, often publishes letters and guides that provide elaboration or clarification of laws and regulations. For example, concern over an increase in outsourcing in 1992 led Administrator Allan Burman to issue extensive guidance to the federal agencies on the classification of their various functions as either inherently or not inherently governmental.

The Secretary of Defense and members of the Secretary's staff (e.g., undersecretaries, assistant secretaries, directors), as the leaders of acquisition in the DoD, also publish numerous orders and directives that constitute acquisition

1413

THE WHITE HOUSE

WASHINGTON

March 10, 1978

Presidential Directive/NSC-33

TO: The Secretary of State
The Secretary of Defense
The Director, Office of Management and Budget
The Director, Arms Control and Disarmament Agency
The Chairman, Joint Chiefs of Staff
The Director of Central Intelligence
The Administrator, National Aeronautics and Space Administration
The Director, Office of Science and Technology Policy

SUBJECT: Arms Control for Anti-satellite (ASAT) Systems (C)

Reference is made to National Security Advisor memorandum, dated September 23, 1977, subject as above.

I direct removal of the restriction, cited in the reference memorandum, on operational or space based testing. The Secretary of Defense is authorized to pursue, for planning purposes, a U.S. ASAT development program encompassing that testing in space or against U.S. objects in space deemed essential to achieve an ASAT capability.

Our future dialogue with the Soviets on Space Arms Control should indicate that we intend to seek an ASAT capability as soon as possible unless they are willing to take very positive actions to preclude such a move on our part.

All other elements of the referenced memorandum remain in effect at this time.

Jimmy Carter

Declassified/Released on 2-11-94
under provisions of E.O. 12356
by D. Van Tassel, National Security Council
F88-195

Fig. 2.1 President Carter's top secret ASAT memo.

policy. In addition to his IPPD-IPT directive mentioned earlier, Secretary Perry in 1994 published a memo that directed the DoD to move away from military unique specifications and standards (MIL-SPECs and MIL-STDs) toward a greater reliance on performance and commercial standards. Such orders and directives are usually formalized and incorporated in the next issuance of internal DoD instructions, such as those described next.

2.6.4 INTERNAL DEPARTMENT OF DEFENSE INSTRUCTIONS AND GUIDANCE

The DoD publishes several policy documents, usually signed by the Secretary or a principal staff member, to guide its internal operations. The most important of these is a set of documents known as the "5000 Series" directives and instructions. These reflect the acquisition goals of the administration, as well as a generalized framework for acquisition program management—including review and oversight within the DoD. The 5000 Series is extremely important for providing specific "how-to" guidelines. For example, it provides a detailed road map of project-management phases and milestones (see Chapter 1) for defense acquisition programs. Although the specifics of this road map might change marginally from administration to administration (as mentioned earlier), the general structure has remained remarkably stable over time (Fig. 2.2).

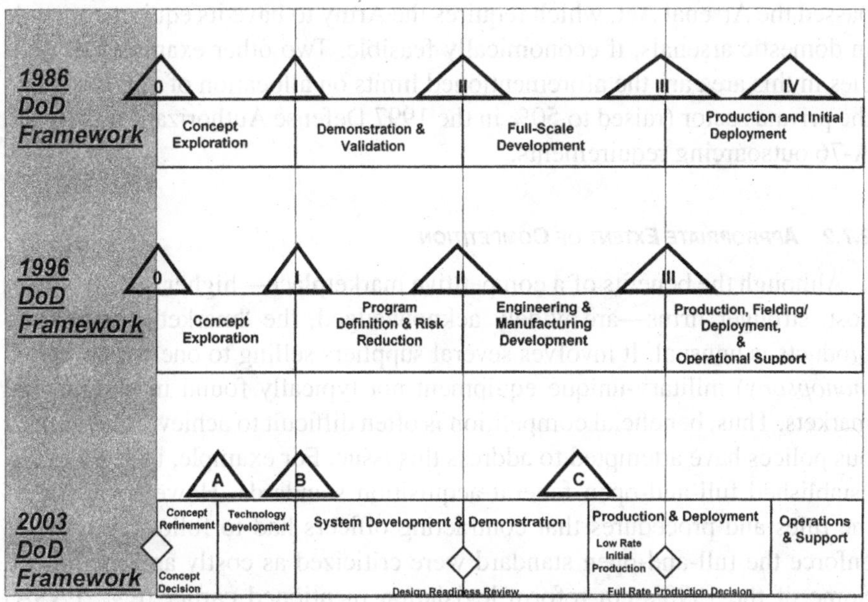

Fig. 2.2 Examples of the changing acquisition process from the 5000 Series.

Because the 5000 Series are internal DoD operating documents, they do not carry the same force as laws or the *FAR/DFARS*. In fact, acquisition managers are encouraged in these documents to tailor their provisions in ways consistent with sound business practices. On the other hand, these documents also contain many statutory and regulatory requirements from public law and the *FAR/DFARS*, which, of course, acquisition managers are obliged to obey.

2.7 IMPORTANT ACQUISITION POLICY AREAS

This section presents several areas toward which acquisition policy has been targeted. This list is by no means exhaustive or prioritized in order of importance, but rather is intended to help the reader appreciate the truly broad spectrum and scope of acquisition policy.

2.7.1 USE OF PUBLIC VS PRIVATE SOURCES

Nations must decide—as matters of defense, industrial, and economic policy—the extent to which they will maintain state-owned assets, such as depots, shipyards, arsenals, and other resources to provide defense equipment and services or will turn to the private sector as a source. Throughout much of its history, the United States maintained significant organic weapons manufacturing capabilities, such as the arsenal at Harpers Ferry. In 1920, Congress passed the Arsenal Act, which requires the Army to have its equipment made in domestic arsenals, if economically feasible. Two other examples of policies in this area are the aforementioned limits on allocation of depot work to the private sector (raised to 50% in the 1997 Defense Authorization Bill) and A-76 outsourcing requirements.

2.7.2 APPROPRIATE EXTENT OF COMPETITION

Although the benefits of a competitive marketplace—higher quality, lower cost, stronger firms—are widely acknowledged, the "market" for defense products is unusual. It involves several suppliers selling to one buyer (called *monopsony*) military-unique equipment not typically found in commercial markets. Thus, beneficial competition is often difficult to achieve, and numerous polices have attempted to address this issue. For example, in 1984 CICA established full-and-open federal acquisition standards. However, many of the rules and procedures that contracting officers had to follow in order to enforce the full-and-open standard were criticized as costly and inefficient. Some of the 1990's-era reform legislation mentioned earlier (e.g., FASA) attempted to address these criticisms by streamlining and simplifying competition and contracting processes.

2.7.3 *APPROPRIATE EXTENT OF REVIEW AND OVERSIGHT*

Acquisition officials seek a proper balance between empowering lower-level officials to make decisions and exercising their own oversight responsibilities. Too much oversight with too many reviews and too many reports creates a burdensome and risk-averse managerial culture. Too little oversight and too few reviews create opportunities for mistakes and errors in judgment to go unnoticed. In the mid-1980s, executive control of acquisition programs intensified because of the horror stories mentioned earlier, and it waned under the "reinventing government" and legislative reform initiatives of the mid-1990s. As noted earlier, internal DoD operating policies have encouraged program managers to tailor acquisition strategies to remove reviews and reports not required by law or regulation. Another example of policy in this area is the executive order signed by President Reagan that mandated compliance with the Packard Commission's recommendations. One of the key features was the creation of a short, streamlined chain of command for acquisition decision making that minimized the number of intermediate managers between a program manager and the program's ultimate decision authority.

2.7.4 *REPORTING TO CONGRESS*

As already mentioned, Congress exercises significant oversight of defense acquisition. One of the principal ways it oversees acquisition is by enacting legislation that requires the DoD to submit reports on a wide variety of acquisition activities. Types of mandatory reports that give the Congress information about progress on specific acquisition programs include *exception* reports (such as Nunn–McCurdy reporting on cost threshold breaches) and *periodic* reports (such as the annually submitted Selected Acquisition Reports), which give cost, schedule, and performance information for major programs. Congress often requires special reporting on programs of interest (see Box 2.3). It also often requires reports on the status of DoD-wide managerial activities and processes. For example, the 2007 Defense Authorization Bill requires the DoD to submit a semi-annual report on the progress of acquisition transformation.

2.7.5 *APPROPRIATE CONTRACTING METHODS*

Ideally, federal contracting should result in the proverbial "win-win" situation, where the government receives a quality product or service on schedule, and the contractor receives fair and reasonable compensation. Much acquisition policy, particularly in the *FAR*, is directed toward achieving this ideal. In cases, though, harmful conditions can arise, such as when the government enforces a fixed-price contract for a developmental type of effort (such as an R&D project with substantial technical risks). Such tactics were actually favored by some officials under the Reagan administration as a way to control costs. However, the case of the failed A-12 Avenger [10] illustrates how this situation can encourage "buy-ins" from contractors, unwarranted optimism by program managers, and other pathological behaviors. Congress severely restricted the use of fixed-price contracts for R&D efforts in 1988.

2.7.6 REQUIRED PROGRAM DESIGN ACTIVITIES

Acquisition program managers must ensure that the designs of their systems address a wide variety of considerations. Some of these requirements are based in law; others are contained in executive directives and guidance documents. Interoperability; human systems integration (HSI); environmental, safety, and occupational health (ESOH); accessibility; and unique identification of items (UID) are just a few examples of these considerations. To assist acquisition managers in understanding and complying with these policy requirements, the DoD publishes helpful guides and other resources (e.g., the *Defense Acquisition Guidebook*).

2.7.7 NATURE AND EXTENT OF TESTING

The testing of defense systems and products provides assessments of their quality and performance. Both contractors and the government conduct tests throughout the acquisition process to determine whether an item's requirements are met. Ideally, sufficient amounts and types of tests are conducted for that purpose. However, because testing consumes time and money, it is often inevitably constrained by program budgets and schedules. Such limits lead to the question, "How much testing is enough?" Historical incidents of inadequate testing have led to two major bodies of legislation regarding testing of weapon systems before they enter full-rate production: first, the requirement to conduct realistic operational testing, and second, the requirement to conduct vulnerability and lethality testing.

2.7.8 ETHICS

Contracting in early American history was so rife with conflicts of interest and favoritism that these evils were not even considered corruption [11]. Among members of Congress, graft was commonplace in their approvals for transcontinental railway contracts, funds, and rights-of-way in the late 1800s [12]. Today, such practices are explicitly prohibited and punishable under criminal law (Box 2.5). Although it is common practice for industry executives to take positions of high responsibility in the DoD, all acquisition officials are expected to follow strict ethical guidelines that establish standards of conduct to promote and uphold the integrity of the acquisition process. For example, federal law (41 U.S.C. 423) prohibits DoD personnel from participating in procurements valued at more than $100,000 when they are seeking employment with any of the bidders.

BOX 2.5 OPERATION ILL WIND ([13], P. 88)

In 1986, the Justice Department launched Operation Ill Wind, the biggest and most successful federal investigation ever conducted of defense procurement fraud. By 1993, 9 government officials, 42 Washington consultants and corporate executives, and 7 companies had been convicted of various crimes. One firm, Unisys, paid $163 million in fines, damages, and forfeitures, plus the cost of the investigation. The highest ranking official involved in the scandal was Melvyn R. Paisley, Assistant Secretary of the Navy. One scheme of Paisley and defense consultant William M. Galvin involved getting the Navy to select Martin Marietta as prime contractor on a research program. They agreed to form a company called Sapphire Systems that would become a subcontractor on the program, though Paisley's financial interest in it would naturally remain concealed. By the end of April 1986, initial funding for the project had been approved, with $900,000 to go to Marietta and $300,000 for Sapphire.

2.7.9 ACQUISITION WORKFORCE PROFESSIONALIZATION

Over the years, acquisition workers have received occasional criticism directed at their competence to handle complex procurements in a manner that protects the public interest. Contracting specialists, for example, have been criticized as less skillful and sophisticated than their industry counterparts, who are supposedly able to negotiate successfully for better contract terms. Also, military program managers have been criticized for being more concerned with their careers than with their programs. As a result, numerous policies, most notably the DAWIA mentioned earlier, have attempted to make the workforce more professional—that is, more trained, better educated, more experienced, and with incentives appropriate for sustained and successful careers in acquisition.

2.8 CONCLUSION

Most in acquisition will never make a single policy during their careers, but every acquisition professional's daily activities are inextricably linked to many different policies. To ignore acquisition's public policy imprint is to ignore one of its distinguishing features. More importantly, ignoring policy is a recipe for failure in acquisition program management, as this chapter has indicated. Subsequent chapters on acquisition's various functional areas (e.g., contracting, logistics, etc.) will highlight and describe in detail the specific policies that are most important in those areas.

REFERENCES

[1] Easton, D., *A Framework for Political Analysis*, Prentice-Hall, Upper Saddle River, NJ, 1965, p. 96.

[2] Lasswell, H., *Politics—Who Gets What, When, How*, Meridian, New York, 1958.

[3] General Accounting Office, "Update on DOD's Purchase of Black Berets," GAO-02-165, Washington, D.C., December 2001.

[4] The Constitution of the United States, Article 1, Sec. 8.

[5] *Federal Acquisition Regulation*, Washington, D.C., March 2005.

[6] "Horror Stories," *National Review*, Vol. 37, No. 17, 6 Sept. 1985, pp. 19, 20.

[7] Gore, A., Letter to President Clinton, "National Performance Review," Washington, D.C., Sept. 7, 1993.

[8] General Accounting Office, "Assessment of Joint Close Support Requirements and Capabilities Is Needed," GAO/NSIAD-96-45, Washington, D.C., 1996, p. 22.

[9] Gansler, J., and Lucyshyn, W., *Implementing Alternative Sourcing Strategies: Four Case Studies*, IBM Center for the Business of Government, Washington, D.C., 2004.

[10] Stevenson, J., *The $5 Billion Misunderstanding: the Collapse of the Navy's A-12 Stealth Bomber Program*, Naval Inst. Press, Annapolis, MD, 2001.

[11] Nagle, J., *A History of Government Contracting*, 2nd ed., George Washington Univ. Press, Washington, D.C., 1999, p. 6.

[12] Bain, D., *Empire Express: Building the First Intercontinental Railroad*, Penguin, New York, 1999, pp. 418–424; 444–445.

[13] Ross, I., "Inside the Biggest Pentagon Scam," *Fortune*, vol. 127, No. 1, 1993, pp. 88–93.

STUDY QUESTIONS

2.1 What is the difference between acquisition in the public and private sectors?

2.2 What are the main reasons that acquisition policy is made?

2.3 Who are the principal sources of acquisition policy, and what forms do their policies take?

2.4 Why does Congress take such an active role in acquisition policy making?

2.5 Other than examples given in this chapter, what are some examples of acquisition policy from the last three years? Why were these policies made? Do they fit in any of the important acquisition policy areas described in the chapter?

Chapter 3

INITIATING DEFENSE ACQUISITION PROJECTS

KEITH F. SNIDER*
NAVAL POSTGRADUATE SCHOOL, MONTEREY, CALIFORNIA

LEARNING OBJECTIVES

- Understand the two principal ways in which acquisition projects begin
- Describe the general process of need development
- Understand the decomposition of a need into operational requirements
- Describe operational requirements and key performance parameters
- Describe the three categories of science and technology (S&T) efforts
- Understand how S&T efforts contribute to acquisition projects
- Describe advanced concept technology demonstrations (ACTDs)

3.1 *INTRODUCTION*

Defense acquisition projects typically begin in response to one of two circumstances: 1) when analyses of the defense environment indicate some gap between what the operational forces have and what they need to accomplish a particular mission or function or 2) when a new technology emerges that provides a new or enhanced operational capability.

This chapter presents an overview of both of these means of initiating a new acquisition project. It discusses processes through which needs and required capabilities are developed, documented, and refined throughout the project life cycle. It also describes the relevance and contributions of science and technology (S&T) activities to the management of defense acquisition programs.

---

*Associate Professor, Graduate School of Business and Public Policy.

3.2 NEED DEVELOPMENT

Figure 3.1 depicts a general process of identifying a required capability to be satisfied by a new defense acquisition project. This process is rational in the sense that the start of an acquisition project is linked to and dependent upon the articulation of higher-level policies, strategies, missions, and objectives.

National security and national defense policies are the purview of the President and his appointed officials and advisors (e.g., Secretary of Defense, National Security Advisor). These top-level policies reflect each administration's particular goals and priorities in ensuring the nation's protection from potential enemies.

National military strategies are developed by senior uniformed leaders and staffs to implement and execute the administration's security and defense policies. These strategies provide more detailed plans and priorities for executing military operations in support of national defense missions and objectives. To execute these strategies and missions, military forces must accomplish various operational functions (e.g., communications, logistics) in different environments. Senior military leaders and staffs periodically assess the ability of forces to accomplish these functions under a range of operational scenarios, contingencies, and threats. This analysis can identify a gap between what the force needs to do and what it is able to do. In some cases, this gap can be closed through some nonmateriel change, for example, organization, tactics,

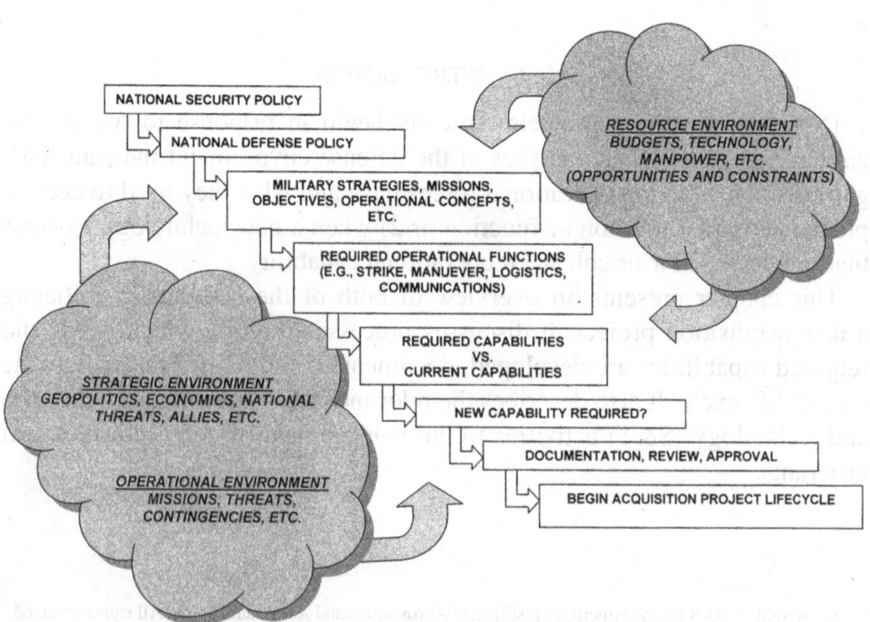

Fig. 3.1 Need development.

doctrine, or training. In other cases, a materiel solution is required to provide the required capability. In all cases, however, the benefits of potential solutions should be weighed against available resources to ensure that only affordable, high-priority programs are pursued (Box 3.1).

BOX 3.1 SEPARATING NEEDS FROM WANTS

Given the current fiscal environment, agencies must separate wants from needs to ensure that programs provide the best return on investments. … Agencies also need to translate their true needs into executable programs by setting realistic and stable requirements, acquiring requisite knowledge as acquisitions proceed through development, and funding programs adequately. However, agencies too often promise capabilities they cannot deliver and proceed to development without adequate knowledge. As a result, programs take significantly longer, cost more than planned, and deliver fewer quantities and different capabilities than promised. Even if more funding were provided, it would not be a solution because wants will usually exceed the funding available [1].

For materiel solutions, the needed capability is formalized in documentation, which then undergoes a process of review, validation, and approval. The level and extent of review depends on the potential importance of the program and the resources to satisfy the need. Hence, the most lengthy and costly programs will be reviewed by the most senior officials.

Much of the analysis and review that results in the documentation of a validated and approved required capability is accomplished by agencies and staffs charged with representing the interests of operational forces in the acquisition process. Consequently, these agencies and staffs are often referred to as "user representatives" or, more simply, the "user."

Once the needed capability is approved through user channels, it is forwarded to acquisition officials for entry into the phases and milestones of the project life cycle.

The identification and development of needs could proceed in different ways. The left side of Fig. 3.2 depicts a "bottom-up" process in which the military services separately generate requirements in response to national security goals. This approach can lead to a number of problems, for example, with duplicative programs and systems that have poor interoperability. For these reasons, in 2003 the DoD instituted a "top-down" process (shown on the right side of Fig. 3.2) designed to ensure that new acquisition programs were initiated in response to jointly identified capabilities, thereby reducing the risk of wasted resources and ineffective systems [2].

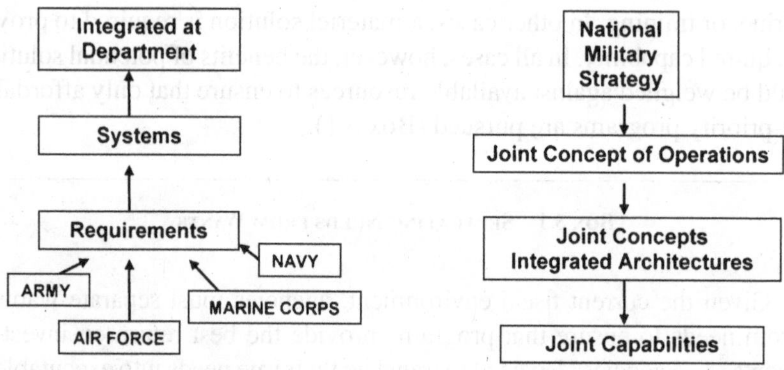

Fig. 3.2 Bottom-up vs top-down requirements process.

3.3 REQUIREMENTS MANAGEMENT THROUGH THE PROJECT LIFE CYCLE

Activities and processes throughout the project life cycle (Fig. 3.3) serve to decompose the approved need into progressively defined, lower-level system, subsystem, and component requirements. Systems engineering (see Chapter 4) accomplishes this decomposition and definition, which results in a system design that can be tested, produced, and fielded to satisfy the need.

Following need approval, a concept development phase ensues for determining the most appropriate system solution to provide the needed capability. This phase often entails parallel study efforts, which analyze various alternatives in terms of their relative advantages and disadvantages. Such analysis is critical for selecting the system design concept most likely to satisfy the need within cost and schedule constraints. Too often, however, the DoD has been criticized for selecting a system solution without appropriate analysis (Box 3.2).

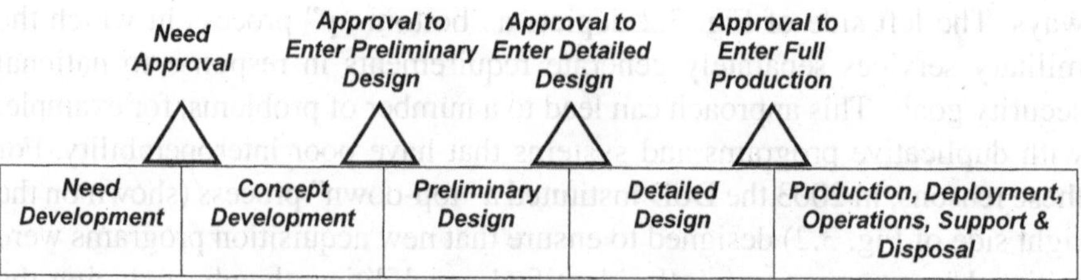

Fig. 3.3 Project life cycle.

BOX 3.2 ANALYSES OF ALTERNATIVES

Though the Navy conducted a formal requirements process and an analysis of other potential solutions, it did so after concluding that the LCS [Littoral Combat Ship] concept was the best option to address challenges of operating U.S. forces in the littorals. Normally, a major acquisition program should include an examination of basic requirements and an analysis of potential solutions before a new system is decided upon. Based on Department of Defense (DOD) reviews of the Navy's analysis and the requirements of revised acquisition guidance, the Navy eventually examined a number of alternative solutions to address littoral capability gaps, such as the extent to which existing fleet assets or joint capabilities could be used. The Navy still concluded that the LCS concept was the best option [3].

A key activity leading up to the preliminary design phase is the translation of the need into a set of operational requirements. Initially, needs are typically stated in very general terms (e.g., "provide deep-strike capability"), and, thus, they do not provide sufficient detail to enable system design. Operational requirements might include parameters such as range, speed, and reliability.

The development of operational requirements closely follows the analysis and development of system solutions to provide the needed capability. Various system solutions for the same need might have different operational requirements. For example, for a deep-attack capability, one might propose two possible system solutions: an attack aircraft system and a surface-to-surface missile system. Clearly, the requirements for speed, survivability, communications, target detection, and many other parameters would differ dramatically between these two alternatives. Operational requirements are usually stated in terms of threshold (minimally acceptable) and objective (desired) values to allow for design tradeoffs during systems engineering.

Development of a concept of operations (e.g., how the system will be employed; how it will be organized, etc.) also accompanies development of operational requirements. To use the preceding example of a deep-attack need, the operational employment and organization of an attack aircraft solution would differ dramatically from that of a surface-to-surface missile solution.

The operational requirements and the operational concept, together with other proposed features of the system solution, contribute to the definition of a system-level specification that will serve as the basis for further design. As described in Chapter 4, this design will be accomplished progressively by systems engineering as the system specification is decomposed into lower-level item and component specifications.

TABLE 3.1 SELECTED KEY PERFORMANCE PARAMETERS FOR THE ADVANCED AMPHIBIOUS ASSAULT VEHICLE [4]

Operational requirement	Threshold value	Objective value
Land speed	69 kph	72 kph
(smooth, level, hard-surfaced roads)		
Reliability	70 h	95 h
(mean time between critical mission failures)		
Carrying capacity	17 Marines	18 Marines

As the design proceeds, the stated operational requirements continue to dictate the design's required level of performance. Obviously, a change in any operational requirement will likely require some change in system design. The most critical, "make or break," operational requirements are called key performance parameters (KPPs); these play an important role in the management of acquisition projects. If the design will not satisfy a KPP, satisfaction of the overall need is at risk. Decision makers at that point must decide among several alternatives, including whether to relax the KPP to a level the design can achieve, to invest more resources in an attempt to satisfy the KPP, or perhaps to terminate the project (see Table 3.1).

Recall from the preceding information that user representatives accomplish the identification, documentation, validation, and approval of a required capability; they also accomplish the development of operational requirements to satisfy that capability. Clearly, an effective interface between the user who develops requirements and the PM who develops systems to meet those requirements is critical for success in acquisition projects.

3.4 *SCIENCE AND TECHNOLOGY IN ACQUISITION PROJECTS*

The benefits of investing in S&T efforts are widely recognized. S&T contributes to advanced systems capabilities that provide military forces with tangible operational benefits and advantages. S&T investments generally fall into the following categories [5]:

- Basic research typically involves long-range research projects that seek to advance the state-of-the-art of fundamental knowledge. Such projects might never have a practical application because they attempt to address basic questions of nature. Basic research necessarily involves a longer-term, more risky investment, but also one with the possibility of a very high payoff.
- Applied research (also called exploratory development) differs from basic research in that it focuses on specific objectives. These research efforts

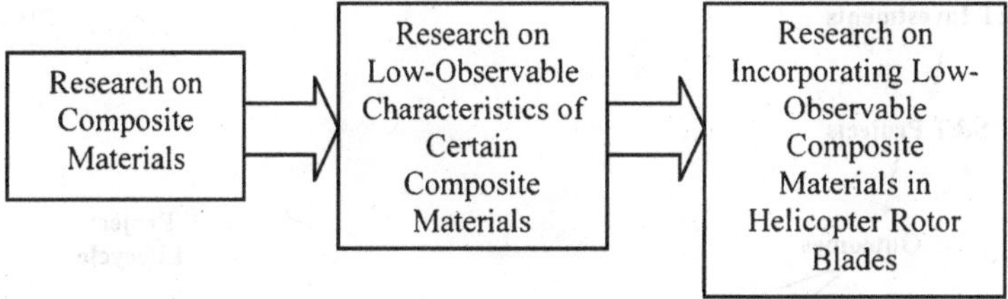

Fig. 3.4 Basic research, applied research, and advanced development.

attempt to bridge the gap between basic research and the initial development of technology for some application. It often entails proof-of-concept experiments that show the practical (though not necessarily military) application of basic research discoveries.

- Advanced-development research projects incorporate the initial development of hardware for use in a defense system. Advanced development focuses on direct applications of scientific and technological advances for military use.

Figure 3.4 gives a notional example of the relationships among these three S&T categories.

S&T efforts are conducted for the U.S. Department of Defense (DoD) by a variety of entities in both the public and private sectors, including universities, DoD and other government labs and R&D centers, industry, and federally funded R&D centers.

S&T efforts are typically not included as part of formally approved acquisition projects because they do not provide a usable end item for fielding. However, they do contribute to the acquisition of defense systems in two major ways (Fig. 3.5):

- S&T activities can illuminate promising and sufficiently mature technologies that would provide such an enhanced operational capability that a new acquisition project is warranted.
- S&T activities can contribute critical and beneficial technologies throughout system design and development, as well as to system modifications throughout its service life.

3.5 Technology Demonstrations

DoD agencies that invest in S&T projects typically conduct demonstrations of the operational utility and benefits of those projects. These advanced technology demonstrations (ATDs) provide wider visibility of the outcomes of S&T projects to user representatives, project/program managers, and others for potential incorporation in acquisition projects.

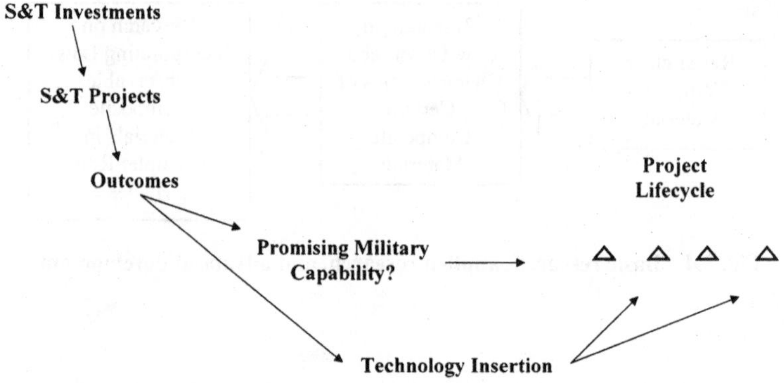

Fig. 3.5 S&T contributions to acquisition projects.

An important development in defense acquisition in the mid-1990s was the advent of ACTDs. The intent of this initiative was to provide a rapid means of placing promising advanced technologies from S&T projects into the hands of the operational forces. Users could then assess the systems' worth under realistic conditions, and those assessments would provide a basis for decisions on whether to let the operational forces retain the capability permanently, transition the technology to a formal acquisition program, return the technology for more S&T work, or terminate the effort. Examples of successful ACTDs include the Global Hawk and Predator unmanned aerial vehicles.

3.6 CONCLUSION

This chapter has described the two principal ways in which acquisition projects are catalyzed: need development and S&T efforts. The manner in which these activities are conducted has major implications for success in any acquisition project, yet the acquisition project manager typically has no authority for and little control over these activities. This indicates how vitally important it is for the project manager to work closely with both user representatives and technologists to ensure that the project has a sound basis in both its requirements and its technologies.

REFERENCES

[1] Government Accountability Office, "Federal Acquisitions and Contracting: Systemic Challenges Need Attention," GAO-07-1098T, Washington, D.C., 2005, p. 1.

[2] Chairman of the Joint Chiefs of Staff, "Joint Capabilities Integration and Development System," Instruction 3170.01F, Washington, D.C., May 2007.

[3] Government Accountability Office, "Plans Need to Allow Enough Time to Demonstrate Capability of First Littoral Combat Ships," GAO-05-255, Washington, D.C., March 2005, p. 3.

[4] U.S. Marine Corps, Combat Development Command, Requirements Div., "Operational Requirements Document (ORD) for the Advanced Amphibious Assault Vehicle (AAAV) (NO, MOB 22.1)," unpublished government document, 1996.

[5] Beason, D., "DOD Science and Technology: Strategy for the Post-Cold War Era," National Defense Univ. Press, Washington, D.C., 1997, http://www.ndu.edu/inss/Press/NDUPress_Books_Titles.htm [retrieved 2 Feb. 2008].

STUDY QUESTIONS

3.1 The phrases "requirements pull" and "technology push" are sometimes used to describe the two ways in which acquisition projects begin. Are these terms appropriate, and why?

3.2 What is the importance of key performance parameters (KPPs)?

3.3 What weapons systems in current use seem to have no existing valid need?

3.4 See Fig. 3.4. Give other examples that illustrate the relationships among the three S&T categories.

3.5 What specific actions can an acquisition project manager take to ensure that both requirements and technology are addressed properly in his/her project?

Chapter 4

SYSTEMS ENGINEERING

THOMAS V. HUYNH* AND KEITH F. SNIDER†
NAVAL POSTGRADUATE SCHOOL, MONTEREY, CALIFORNIA

LEARNING OBJECTIVES

- Appreciate the value of systems-engineering efforts in project management
- Define systems and systems engineering
- Describe the systems-engineering process: its inputs and outputs, its steps, and system analysis and control activities
- Describe the purpose and importance of configuration management
- Identify the various types of specifications and baselines
- Understand the importance of systems integration
- Know the purpose of design reviews in the systems-engineering process
- Describe the systems-engineering master plan (SEMP) and master schedule (SEMS)
- Describe the content and use of the work breakdown structure (WBS)
- Know the purpose of technical performance measures (TPMs)
- Understand the relationship between systems-engineering activities and the phases and milestones of the project life cycle
- Describe systems of systems, their importance, and challenges in their management

---

*Associate Professor, Department of Systems Engineering.
†Associate Professor, Graduate School of Business and Public Policy.

- Understand characteristics of open systems and open architectures
- Describe modular open system approaches (MOSA)

4.1 INTRODUCTION

This chapter introduces the principles and practices of systems engineering as it pertains to acquisition. Systems engineering is the principal technical discipline that guides the development and production of systems that satisfy user requirements within cost and schedule constraints; thus, its criticality in ensuring successful acquisition outcomes cannot be overstated.

The chapter begins by establishing the value that systems engineering contributes to defense acquisition projects. It then defines systems and systems engineering. A large portion of the chapter is devoted to the systems-engineering process, its steps and activities, and its connections to the acquisition project life cycle. The chapter concludes with discussions of systems of systems, open systems, and open architectures.

In this brief space, we can only sketch the outlines of systems engineering; we do not claim that studying this chapter will qualify anyone to work as a systems engineer. Our purpose is rather to acquaint the reader with a description of what systems engineers do and to impart an appreciation for their importance in acquisition projects.

The majority of systems-engineering activities in defense acquisition projects are accomplished by private industry, specifically, by firms under contract to develop and produce systems. Government project-management offices and other acquisition agencies can perform systems-engineering tasks in the early need and concept development phases of the project life cycle. During the design, development, and production phases, however, government personnel typically provide leadership through the review and oversight of contractor systems-engineering efforts. Thus, although government personnel generally need not be experts in systems engineering, they must either know enough or have sufficient systems-engineering support in order to perform those review and oversight functions properly.

4.2 VALUE OF SYSTEMS ENGINEERING IN DEFENSE SYSTEMS ACQUISITION

The acquisition of large-scale, complex systems calls for effective systems-engineering practices intertwined with effective management practices. Unfortunately, many acquisition pwrograms have failed because of a lack of organizational support for systems engineering, unsound systems-engineering practices, and poor systems-engineering management (Box 4.1).

BOX 4.1 INADEQUATE SYSTEMS ENGINEERING IN THE ARMY'S CRUSADER PROGRAM ([1] PP. 42, 43)

The Army's Crusader artillery vehicle is a case in which systems engineering was not performed by the product developer, to a large extent, until after the acquisition program had been launched. The development program was launched based on a notional design done by the Army's program office, not extensive systems engineering done by the contractor that would design and build the product. Key to this notional design was the use of a liquid propellant—new technology—for firing weapon projectiles. The Army assessed various aspects of the risk of developing the liquid propellant technology and integrating it into the weapon system between low and moderately high. Nevertheless, on the basis of the notional design, the Army committed to launching product development. After the program was launched in 1994, the product developer was awarded a contract to develop the Crusader. According to a program official, it took 2 years of systems engineering to determine if the requirements were feasible given established cost and schedule targets. In 1996, the product developer determined that the liquid propellant technology was high risk in all aspects and that it would cost an additional $500 million to develop.

Systems engineering clearly brings value to acquisition programs. Results from a recent study indicated that systems engineering can be effective in controlling cost overruns and schedule slippage (Fig. 4.1). In light of recent unsuccessful acquisition outcomes, and considering systems-engineering's positive contributions, recent U.S. acquisition policy seeks to revitalize systems-engineering efforts in acquisition program management [3].

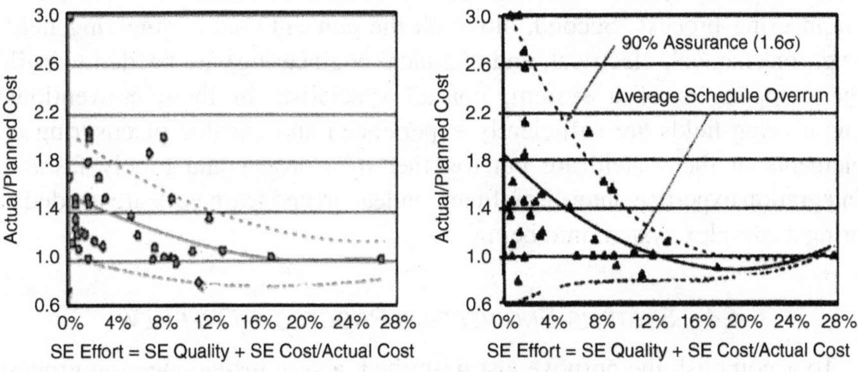

Fig. 4.1 Cost and schedule overruns correlated with systems-engineering effort [2].

4.3 SYSTEMS AND SYSTEMS ENGINEERING

A system can be defined as

> [A] construct or collection of different elements that together produce results not obtainable by the elements alone. The elements, or parts, can include people, hardware, software, facilities, policies, and documents; that is, all things required to produce systems-level results. The results include system level qualities, properties, characteristics, functions, behavior and performance. The value added by the system as a whole, beyond that contributed independently by the parts, is primarily created by the relationship among the parts; that is, how they are interconnected [4].

A shorthand definition of a system involves a quadruplet: components, attributes, relationships, and purpose. Components constitute the system. Attributes are the properties of the components. Relationships are the links between the components and their attributes. A conglomeration of items would not constitute a system unless it displays unity (resulting from integration), functional relationship (output of a function is input to others), and a useful purpose (to satisfy a need or needs). Examples of large and complex systems include the space shuttle, banking systems, information networks, radar tracking systems, chemical plants, and satellites.

The purpose of systems engineering as a discipline is to bring a system into being that meets the requirements for which it is intended within cost and schedule constraints. Systems engineering employs an interdisciplinary engineering process through a top-down approach that views the system as a whole. It addresses system design and development, production and/or construction, distribution, operation, maintenance and support, retirement, phase-out, and disposal. Systems engineering involves the use of appropriate tools, methodologies, and management principles in a synergetic manner to ensure that all design objectives are achieved effectively and efficiently.

Why is systems engineering needed? First, the development and production of a quality system within cost and schedule constraints requires a quality engineering process. Second, although the conventional engineering fields (e.g., mechanical, electrical, and chemical engineering) are needed to build the components of a system, not all specialists in these conventional engineering fields are sufficiently experienced and capable of ensuring all elements of the system are put together in a proper and timely fashion. Integration expertise, interdisciplinary endeavor, and teamwork are needed to bring a complex system into being.

4.4 SYSTEMS-ENGINEERING PROCESS—OVERVIEW

To accomplish the purpose just described, a systems-engineering process (SEP) is employed. The SEP is an "iterative process that expands on the common sense strategy of 1) understanding a problem before you attempt to solve it, 2) examining alternative solutions, and 3) verifying that the selected solution is correct before continuing the definition activities or proceeding to the next problem" ([5], p. 16).

The SEP applies throughout the project life cycle to all activities associated with product and process development, testing, manufacturing, training, deployment, operation, support, and disposal ([5], p. 11).

Several different SEP models exist, but, in essence, each model describes a problem-solving process that identifies and evolves a system's products and processes. This chapter will explore one widely used model, the Institute of Electrical and Electronics Engineers (IEEE) model shown in Fig. 4.2.

The iterative nature of the SEP (Fig. 4.2) means that it begins at the top—the system level—to generate a description of system-level product and process design. Then the SEP is repeated throughout the project life cycle at lower-than-system levels to generate more detailed descriptions of subsystems and components. The final output is a complete description of the desired system. Further, the SEP is an evolutionary process in that it accommodates feedback (from the results of tests, tradeoff analyses, etc.) to improve and refine the system design.

The steps and features of the SEP are briefly covered in the following discussion adapted from ([6], pp. 37–66).

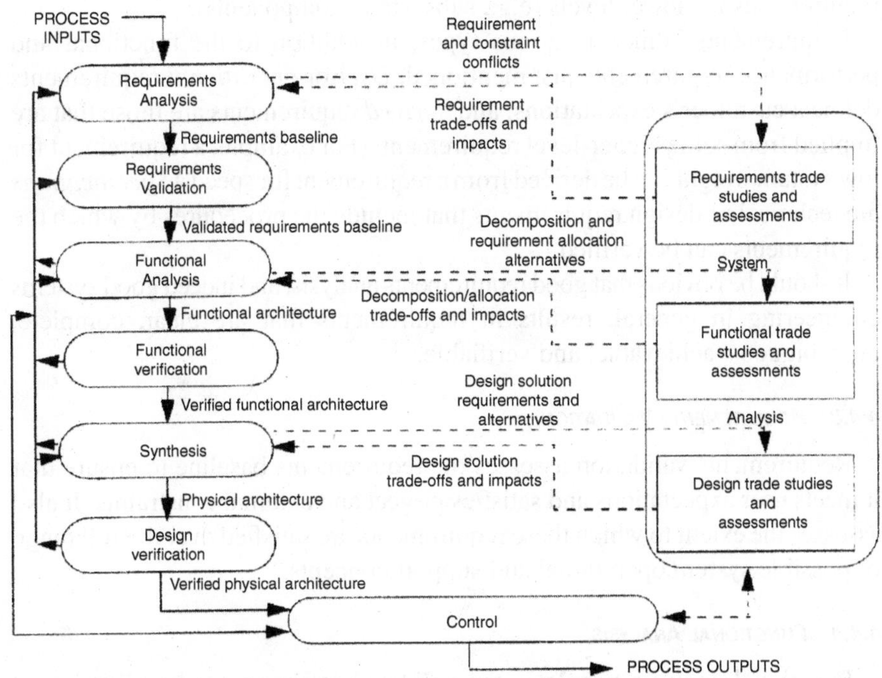

Fig. 4.2 Systems-engineering process ([6], p. 12).

4.4.1 PROCESS INPUTS

Inputs to the SEP consist mainly of the user's needs, objectives, and requirements. They also include other project objectives, as well as opportunities and constraints from the external environment (e.g., available technology). Among the numerous sources of inputs are user statements of needs and requirements, program plans, design documents, system threat documents, technology assessments, and acquisition policies. Because the SEP is iterative, outputs from one iteration become inputs for the next lower-level iteration.

4.4.2 REQUIREMENTS ANALYSIS

Requirements analysis establishes what the system must accomplish (*functional requirements*), how well it must perform (*performance requirements*), the environments in which it must operate, and any other requirements and constraints that might affect system design. SEP inputs—stakeholder expectations, project and external constraints, and higher-level system requirements—are assessed and analyzed to identify cost, schedule, and performance risks, to define functional and performance requirements, and to identify any conflicts. Requirements are documented in a baseline that defines the problem to be solved, thus guiding the remaining activities of the SEP. For each iteration of the SEP, previously defined requirements for higher levels of the system architecture are refined, resulting in progressive definitions of requirements for lower levels (e.g., subsystem, components).

Requirements come in various types, in addition to the functional and performance requirements just mentioned. *User* or *operational* requirements define a customer's expectations, and *derived* requirements are those that are implied from some higher-level requirement. (For example, a requirement for low system weight can be derived from a requirement for speed.) *Specifications* are technical or design requirements that include the procedures by which the requirements can be verified.

It should be obvious that good requirements analysis, and indeed good systems engineering in general, results in requirements that are clear, complete, unambiguous, achievable, and verifiable.

4.4.3 REQUIREMENTS VALIDATION

Requirements validation assesses the requirements baseline to ensure that it meets user expectations and satisfies project and external constraints. It also assesses the extent to which these requirements are satisfied over the full range of possible system operational and support concepts.

4.4.4 FUNCTIONAL ANALYSIS

Functional analysis translates the validated requirements baseline into a *functional architecture*. It defines *functions* (actions or behaviors) necessary

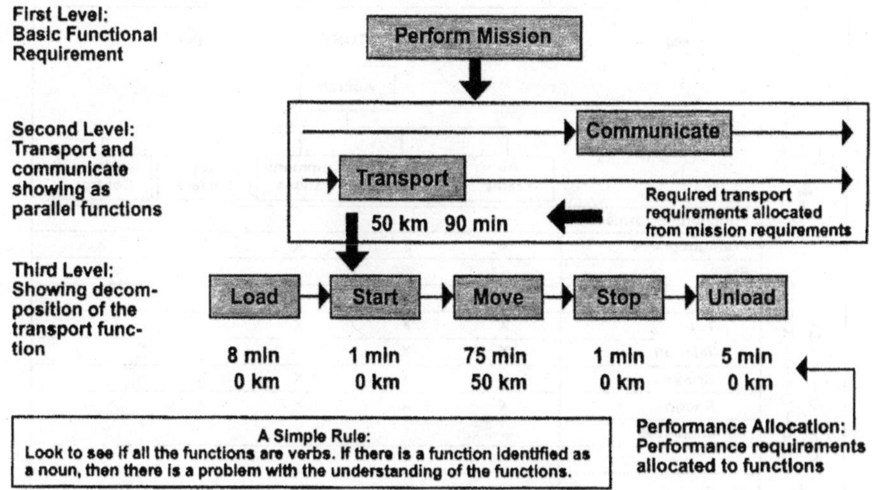

Fig. 4.3 Functional architecture ([7], p. 47).

to accomplish those requirements and decomposes both the *functional requirements* (what must be done) and the associated *performance requirements* (how well it must be done) into lower-level functions. The functional architecture describes the arrangement and sequencing of functions resulting from this decomposition from higher- to lower-level functions. Figure 4.3 depicts a simple example of a functional architecture.

4.4.5 FUNCTIONAL VERIFICATION

Functional verification assesses the adequacy and completeness of the functional architecture to satisfy the validated requirements.

4.4.6 SYNTHESIS

Synthesis defines a design solution to satisfy the requirements of the verified functional architecture. It translates the functional architecture into a *physical* (or design) *architecture* of system elements, their decomposition, interfaces (internal and external), and design constraints. Synthesis involves selecting a preferred design solution from a set of alternatives—taking into account cost, schedule, and performance implications. Figure 4.4 shows a notional translation of a functional architecture to an aircraft physical architecture. Note the allocation of various functions (including how well those functions must be performed) to various physical parts of the aircraft.

4.4.7 DESIGN VERIFICATION

Design verification ensures that requirements of the physical architecture are traceable to the verified functional architecture and that the design satisfies

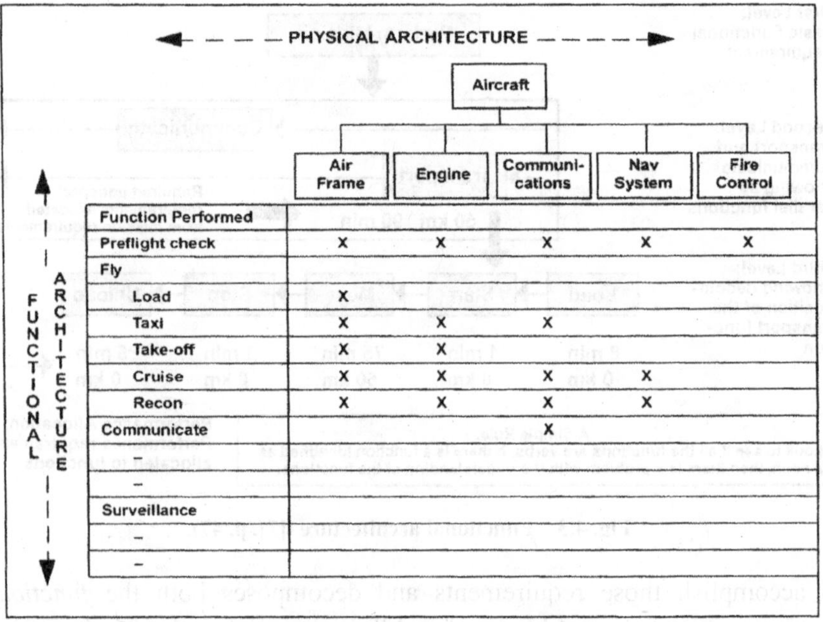

Fig. 4.4 Physical architecture ([7], p. 59).

the validated requirements baseline. Verification can be accomplished through a variety of means, including analysis, inspection, demonstration, and test (see Chapter 7).

4.4.8 SYSTEMS ANALYSIS

Systems analysis activities support the SEP steps just described by 1) resolving conflicts (e.g., competing requirements during requirements analysis), 2) decomposing and allocating functional and performance requirements during functional analysis, 3) evaluating alternative design solutions and selecting the best design solution during synthesis, 4) assessing system effectiveness, and 5) managing risks throughout. Through tradeoff analyses and other activities, systems analysis provides a quantitative basis for a balanced set of requirements and a balanced design.

4.4.9 PROCESS OUTPUTS

SEP outputs are data that describe the configuration of the system and the processes for its development, production, deployment, and support. The level of output specificity depends on the level of development. Outputs become increasingly detailed as system development progresses. As mentioned earlier, as each SEP iteration is completed, outputs form the input for succeeding SEP iterations.

4.4.10 CONTROL

Control activities serve to manage and document the activities of the SEP. They provide complete and current status updates of SEP activities and results; plans for and inputs to future SEP iterations; information for production, test, and support; and information for decision makers at technical and management reviews. Control activities include risk management (see Chapter 8) and other management tasks such as those listed next.

CONFIGURATION MANAGEMENT. Configuration management (CM) refers to activities that control the development of a product or system by ensuring consistency between the design and the finished product. CM controls the progression of an item's configuration (i.e., its functional and physical characteristics, along with its interfaces with other items or systems) through successive iterations of the SEP. CM also controls changes to requirements and design throughout the project's life cycle.

A system's configuration is usually expressed in various levels of *specifications* and *baselines*. Recall from the preceding information that a *specification* is a technical or design requirement that includes the procedures for verifying that the requirement is met. A *baseline* is a collection of specifications that defines a particular level of system development. Table 4.1 defines different levels of specifications and baselines commonly used in defense acquisition project management. The system development contractor will typically develop the baselines, initially in draft form, and then manage the baselines in system design and development. At some point—usually during the design reviews discussed next—the government project-management office will approve and accept "ownership" of the baselines.

A complete description of what is necessary to produce a system (i.e., all specifications, processes, verification procedures, etc.) is given in a *technical data package* (TDP). Obviously, completion of a TDP will not be possible until a system has successfully entered the production phase of the project life cycle.

TABLE 4.1 SPECIFICATIONS AND BASELINES

Specification	Content	Baseline
System spec	Defines mission/technical performance requirements	Functional
Item performance spec	Defines performance characteristics of subsystems and other major components	Allocated ("design to")
Item detail spec	Defines form, fit, function, performance, and test requirements	—
Process spec	Defines process performed during fabrication	—
Material spec	Defines production material used in fabrication	Product ("build to")

CM helps ensure that only one approved version of an item's design exists and that all items produced and fielded conform to that design. Poor or inadequate CM can lead to difficulties in a number of different areas, for example, in manufacturing and in providing support for fielded systems.

CM includes several related efforts. *Configuration identification* provides for the documentation and recording of baselines, specifications, and other CM-related information. *Configuration control* provides for a disciplined process for controlling changes to established configurations. Most projects use a configuration control board for this purpose. *Configuration status accounting* relies heavily on data management to collect, record, and disseminate CM-related information. *Configuration audits* verify that an item conforms to its configuration documentation. Two commonly occurring audits are the functional configuration audit (FCA) and the physical configuration audit (PCA), which verify those two aspects of a system's characteristics, respectively.

INTERFACE MANAGEMENT. Interface management recognizes the importance of and addresses systems integration—one of the most challenging activities in systems engineering (Box 4.2). Systems integration requires a systems engineer to progressively (successively) combine the components and testing of the resulting system to ensure that requirements are met. Up to 40% of the development effort of large and complex systems can be spent in systems

BOX 4.2 INTEGRATION CHALLENGES IN THE ARMY'S DIVAD PROGRAM [9]

System integration problems seemed to plague even relatively simple combinations of reasonably well-understood components. The Army's DIVAD (division air defense) system mated the old M-48 tank chassis with the F-16's APG-68 radar and a production model German gun system. Not a difficult combination, one might think, and on that assumption that Army's program office sought a relatively rapid move to production. Indeed, the Army signed a fixed-price final development and production contract with Ford Aerospace, the winning contractor, which raised the costs of achieving contractual test and production milestones. Yet the Army was forced to violate those milestones, as DIVAD encountered an array of technical and operational test difficulties that ultimately led to its cancellation in 1985. Not all of DIVAD's problems stemmed from systems integration; getting an aircraft radar to work in the ground environment proved to be more challenging than expected. But integrating DIVAD's three basic components contributed to the system's problems and ultimate cancellation.

integration, mostly in system testing [8]. Hardware and software integration has been recognized as a significant challenge in systems integration.

Interface management is a key element in systems integration. It encompasses physical, logical, and human–system interfaces. A successful systems-integration effort necessarily requires designing and managing interfaces among the components or subsystems of the system. Interface management also requires a systems engineer to verify and validate the interfaces to confirm proper linkages exist among the components and to correct information flow across those interfaces.

Systems-Engineering Technical Management. This management task includes several activities, tools, and techniques for managing the SEP:

Design reviews: Formal design reviews are conducted throughout system development (typically at the completion of each SEP iteration) to assess design progress and maturity, as well as to assess program technical risks. Examples of design reviews include system requirements review (SRR), system functional review (SFR), preliminary design review (PDR), and critical design review (CDR).

Systems-engineering management plan (*SEMP*): The SEMP is a plan for managing the systems-engineering effort. It identifies the items that are to be developed, delivered, integrated, installed, verified, and supported. It identifies when these tasks will be done, who will do them, and how the products will be accepted and managed. Finally, it defines the technical processes to be used to produce each of the project's products. Blanchard and Fabrycky [10] provide a SEMP checklist:

- Does the plan address the overall system life cycle and its phases and activities?
- Does the plan describe the systems-engineering process?
- Does the plan properly integrate the different engineering specialties involved in the system or product design process?
- Does the SEMP integrate other plans, such as the reliability program plan, maintainability program plan, safety engineering plan, and integrated logistics support plan?
- Are program tasks, organizational responsibilities, the WBS and schedule requirements, cost estimates, and program monitoring-and-control functions included?
- Have formal design reviews been planned?

Systems-engineering master schedule (*SEMS*): The SEMS captures all major systems-engineering activities and milestones, as well as program milestones. These activities and milestones must be merged and synchronized in an iterative fashion to produce the SEMS.

Work breakdown structure (*WBS*): Development of a WBS (Fig. 4.5) follows the generation of a physical architecture in the SEP. The WBS

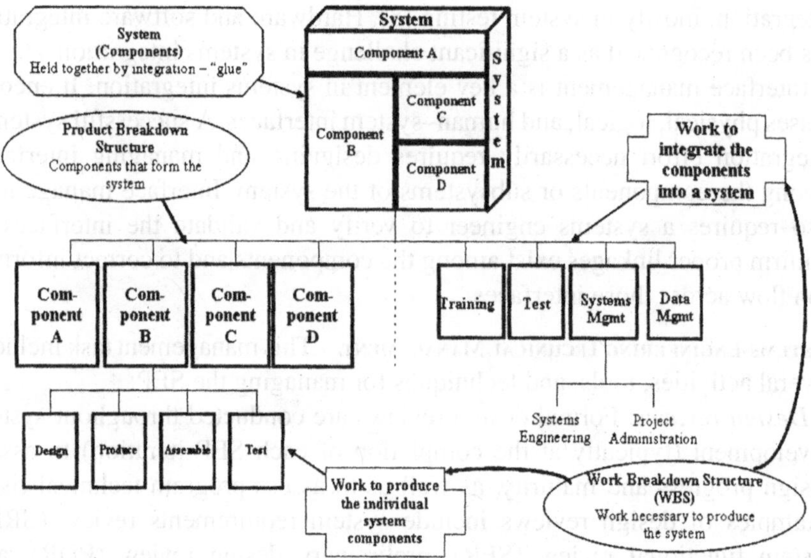

Fig. 4.5 Work breakdown structure (adapted from [11]).

captures the entire scope of work necessary to produce the system. This includes not only the work to produce individual system components (e.g., design, produce, assemble, and test), but also the work to integrate the components into the system (e.g., systems management, training, test, data management, etc.). Because the WBS captures the work necessary to develop and produce a system, it is an excellent tool for project planning, budgeting, scheduling, contracting [e.g., in developing a statement of work (SOW)], risk management, and many other activities.

Technical performance measures (*TPMs*): TPMs are key attributes and/or characteristics that are critical to system performance. A system's progress toward achieving those attributes and characteristics indicates its progress toward satisfying system requirements. Examples of TPMs might be component weight, reliability, throughput, and power output.

4.5 Systems Engineering in the Project Life Cycle

The discussion to this point has focused on systems engineering as a discipline for systems development. It is important, however, to understand how systems engineering fits in and contributes to management of defense acquisition projects. Figure 4.6 gives a simplified view of systems-engineering activities in a notional acquisition project. Note that the SEP is accomplished multiple times throughout the phases of system development and design; the number of SEP iterations will depend upon each program's particular situation and requirements.

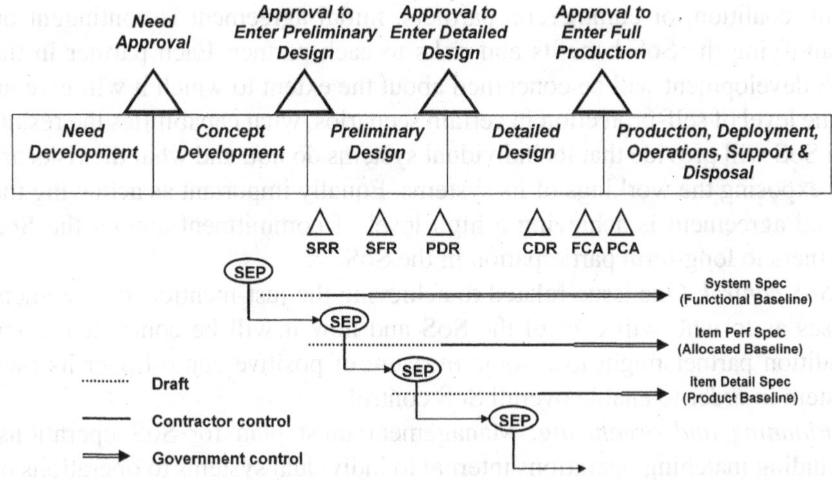

Fig. 4.6 Systems-engineering activities in the project life cycle.

The technical activities of systems engineering will support the managerial decisions and processes of the project life cycle. Likewise, programmatic decisions at the milestones of the project life cycle should rely on information provided by systems engineering regarding the progress of system design and the various risks involved. For example, the timing of the project milestones should align to the greatest extent possible with the timing of appropriate systems-engineering design reviews and audits (e.g., SRR, PDR, PCA).

4.6 SYSTEMS OF SYSTEMS

Recently, there has been increased interest and emphasis on *systems of systems* (SoS) and SoS engineering. Much of this current interest in SoS is focused on the desire to integrate military systems to achieve new capabilities that would be unavailable through any of the existing single individual systems.

Military SoS result from combining separately developed systems, usually operated by separate organizations and usually geographically dispersed, into a larger system, such as a joint or a coalition system. Thus, the focus here considers an SoS as a conglomeration of existing stand-alone systems and also to-be-developed systems that are integrated and interoperable with each other.

Management of SoS projects involves some unique issues [12]:

Initial agreement and commitment: In developing an SoS, decision makers must initially agree that the SoS meets some objective and that this objective is desirable. If the proposed SoS is within the purview of one organization, such as the U.S. DoD, development decisions can be directed from top down. In this case, agreement is not an issue. However, if the proposed SoS involves joint, coalition, or commercial partners, initial agreement is contingent on quantifying the SoS benefits and risks to each partner. Each partner in the SoS development will be concerned about the extent to which it will give up some level of self-protection in certain scenarios, what capabilities the resulting SoS will provide that its individual systems do not, and what the risks are for exposing the workings of its systems. Equally important as achieving the initial agreement is achieving a high level of commitment among the SoS partners to long-term participation in the SoS.

SoS control: One issue related to achieving the just-mentioned agreement arises as to who will control the SoS and how it will be controlled. Each coalition partner might lose some measure of positive control over its own system in order to enable overall SoS control.

Planning and organizing: Management must plan for SoS operations, including matching operations internal to individual systems to operations of external systems. Processes that interface with SoS processes must be established and monitored.

Directing: Clear, concise, and complete SoS communications channels must be established. Measures of successful SoS operations need to be identified.

Staffing: Management must identify the skills required for SoS operations, acquire personnel with the proper skills, and train personnel in SoS operations. These skills can include, for example, new types of communications network operations. By analyzing interfaces, management will identify new tasks that need to be accomplished in order to achieve successful SoS operations. These tasks can be matched to personnel capabilities identified, gaps identified, and requirements for training or hiring or critical skills identified.

Team building: Personnel will be operating in new environments, and internal teamwork and external teamwork must be established through team-building efforts.

Reporting: Management must identify metrics to be reviewed and evaluated for SoS operations; it also must develop measurement and collection plans.

Classified and proprietary design information: The sharing of classified and/or proprietary information is a major SoS management challenge. Commercial firms will have to reap substantial benefits before they will allow access to proprietary information. Military organizations likewise would have to obtain benefits far in excess of potential losses that they might incur through the release of classified information. A corresponding issue is determining what sensitive information needs to be shared in order to develop an SoS.

SoS testing: Each coalition partner needs to be assured that the SoS is tested in a manner that resolves any of its concerns about operational behavior.

Emergent behavior: Finally, the SoS might exhibit emergent behavior that has not been anticipated by any of the partners.

4.7 OPEN SYSTEMS

The U.S. DoD no longer propels and sets pace for technology development. The private sector now develops technologies at a rapid rate and with short cycle times. To field systems quickly and affordably, the U.S. DoD attempts to leverage commercially available technologies and private-sector practices. Effective use of the commercially available products and technologies requires a paradigm shift from proprietary to public architectures, from stove-piped systems to open systems.

An *open system* is one that exchanges energy, material, and information with its environment on a continuing basis, through open (well-defined, widely used, nonproprietary) protocols, standards, languages, and data formats. All aspects of system interfaces in an open system should be well-defined and should, with minimal impact on the system, facilitate system upgrading and the addition of new capabilities. An open system is designed by consensus, using a collection of interacting and integrated software, hardware, human components, and easily accessible standards. In contrast, a *closed system* is one that is designed by a single organizational entity for a single program or small number of programs and in which the standards, protocols, data formats, and languages are unavailable or only partially available [13].

4.7.1 OPEN ARCHITECTURES

An *open architecture* implements an open system, as it uses off-the-shelf components and publicly available standards that allow the system to be connected to and interoperate with diverse devices, programs, and systems. A *closed* or *proprietary architecture* makes it difficult to connect various systems. An open architecture is a public architecture that facilitates add-on designs and duplications. For example, Linux is based on an open architecture, with source code available for free, whereas DOS, Windows, and the Macintosh architecture and operating system are based on closed architectures.

4.7.2 MODULAR OPEN-SYSTEMS APPROACH

Modular open-systems approach (MOSA) is the name for the DoD's focus on using integrated business and technical strategies that employ open systems and open architectures. As a business strategy, MOSA, which promotes the use of commercial products from multiple sources, should enable the defense systems acquisition community to quickly and affordably build, upgrade, or support systems. As a technical strategy, MOSA emphasizes modular system design, well-defined interfaces, and support by established industry standards for key interfaces, to the greatest extent possible. MOSA should enable several benefits: 1) joint integrated architecture and interoperability; 2) reduced system acquisition cycle time and total ownership cost; 3) easier insertion of cutting-edge technology; and 4) commonality and reuse of components among systems [14].

4.8 Conclusion

The current defense environment is one of constant and fast-paced change. Such an environment places a premium on systems that can readily adapt to new requirements. Although our forces today must accomplish an unprecedented array of missions, the rapid expansion of technologies presents them with new ways to accomplish those missions. Such trends drive the demand for enhanced war-fighting capabilities, and as the capabilities of defense systems grow, so will their complexity. Yet, demands for affordability and responsiveness will remain. Only through the application of sound systems- engineering practices can such highly complex and capable systems be produced within cost and schedule constraints. It is safe to say that no acquisition project manager can expect to achieve success without placing the highest priority on systems engineering.

References

[1] General Accounting Office, "Better Matching of Needs and Resources Will Lead to Better Weapon System Outcomes," GAO-01-288, Washington, D.C., March 2001.

[2] Honour, E., "Understanding the Value of Systems Engineering." *Proceedings of the 14th Annual INCOSE International Symposium*, Toulouse, France, June 2004, p. 8.

[3] Department of Defense, "The Defense Acquisition System," Directive 5000.1, Washington, D.C., 12 May, 2003, p. 11.

[4] Rechtin, E., *Systems Architecting of Organizations: Why Eagles Can't Swim*, CRC Press, Boca Raton, FL, 2000, p. 4.

[5] International Council on Systems Engineering (INCOSE), *Systems Engineering Handbook*, ver. 3, INCOSE-TP-2003-002-03, Seattle, WA, June 2006.

[6] IEEE Computer Society, *IEEE Standard for Application and Management of the Systems Engineering Process*, IEEE Std 1220TM-2005, Inst. of Electrical and Electronics Engineers, New York, Sept. 2005.

[7] Defense Systems Management College, *Systems Engineering Fundamentals*, Defense Acquisition Univ., Fort Belvoir, VA, Jan. 2001, p. 47.

[8] Sommerville, I., "System Integration," Computing Dept., Lancaster Univ., Lancaster, UK, 1998, http://www.comp.lancs.ac.uk/computing/resources/Ians/SE5/syseng/size [retrieved September 2003].

[9] McNaugher, T. L., "Joint Strike Fighter (JSF) Acquisition Reform: Will It Fly?," Statement Prepared for the House Committee on Government Reform, Subcommittee on National Security, Veterans Affairs, and International Relations Hearings, Washington, D.C., 10 May, 2000.

[10] Blanchard, B. S., and Fabrycky, W., *Systems Engineering and Analysis*, 4th ed., Prentice-Hall, Upper Saddle River, NJ, 2005, p. 662.

[11] Forsberg, K., Mooz, H., and Cotterman, H., *Visualizing Project Management*, 3rd ed., Wiley, New York, 2005, p. 205.

[12] Osmundson, J., Huynh, T., and Langford, G., "System of System Management Issues," *Proceedings of the Asia-Pacific Systems Engineering Conference*, Singapore, March 2007, p. 92.

[13] Azani, C. H., "The Test and Evaluation Challenges of Following an Open System Strategy," *ITEA Journal*, Vol. 22, No. 3, 2001, pp. 23–28.
[14] Department of Defense, *Program Manager's Guide: A Modular Open Systems Approach to Acquisition*, ver. 2, Open Systems Joint Task Force, Washington, D.C., Sept. 2004.

STUDY QUESTIONS

4.1 Give examples of complex systems, other than those mentioned in the text.

4.2 Explain the differences between project management (as described in Chapter 1) and systems engineering. How are they related?

4.3 How do the inputs to the SEP change throughout the project life cycle? The outputs?

4.4 To what extent is systems engineering relevant in the acquisition of a commercial off-the-shelf (COTS) item?

4.5 Assume that you have just been assigned as the project manager for the acquisition of a new defense system and that the system has both hardware and software components. What specific steps would you take to ensure that proper system integration occurs from the beginning and on a daily basis?

4.6 What information would you expect to be presented at a design review?

4.7 What managerial actions can help ensure that architectures and systems are designed to be "open"?

Chapter 5

SOFTWARE PROJECT MANAGEMENT

JOHN S. OSMUNDSON*
NAVAL POSTGRADUATE SCHOOL, MONTEREY, CALIFORNIA

LEARNING OBJECTIVES

- Understand the importance of software project management
- Understand the challenges of managing software projects
- Understand the influence that requirements analysis has on the success or failure of software projects
- Learn the differences and applicability of various software development models, including waterfall, spiral, and agile development models
- Be exposed to how key features of the unified modeling language (UML) aid in the successful development of software projects
- Be cognizant of the challenges of software project cost estimation and approaches to meet those challenges
- Learn fundamentals of monitoring software project progress
- Understand the importance of configuration management, testing, and maintenance
- Understand basic risk-reward tradeoffs of software reuse and use of commercial off-the-shelf (COTS) software

5.1 *INTRODUCTION*

This chapter discusses the importance and challenges of software projects. Key factors for successful software project management are summarized. The first key factor is good requirements analysis. This is tied to the choice of an appropriate project-development model (several of which are discussed), as well software development standards and the Software Engineering Institutes Capability Maturity Model (CMM). Next, approaches to software

*Research Associate Professor, Dept. of Information Sciences.

project estimation and planning are discussed, followed by project monitoring, metrics, quality assurance, configuration management, testing, software maintenance. This chapter concludes with a section on the risks and rewards of software reuse and use of COTS software.

5.2 Importance of Software Project Management

There is evidence that software is the most important of DoD acquisitions, either as a stand-alone system or as a component of an embedded system. The trend in embedded systems is to require more and more functionality to be performed in software as opposed to hardware. Table 5.1 shows this trend for fighter aircraft avionics. The result is that for embedded systems the cost of software (including maintenance costs) has far outstripped the relative cost of hardware. This trend, over the past 50 plus years, is shown on Fig. 5.1 (data available online at http://software.ssu.ac.kr/SE_page/06_class1.ppt#13).

At the same time, there is abundant evidence that there are difficulties in managing the development of software systems. In a 1979 report [2], the General Accounting Office (GAO) found that more than 50% of DoD contracts for software developments had cost overruns; more than 60% of contracts had schedule overruns; more than 45% of software (SW) could not be used; and more than 29% of SW was never delivered. Less than 2% of SW was usable as delivered. The term "software crisis" was coined to describe the abysmal state of software development.

More recent data from SPAWAR [3] indicate that DoD software developments are still experiencing poor results: 53% of all software projects cost nearly 90% over the original estimates, 42% of the original proposed features and functions are implemented in the final product, and 31% of all software projects are cancelled prior to final delivery.

This rate of failure has also been found by the Standish group [4–7] over a period of 10 years. Figure 5.2 shows the percentage of *successful* projects—defined as those projects that were on time, on budget, and delivered the

Table 5.1 Software Provides Increasing Functionality with Time [1]

Aircraft	Year	% of Functions performed in software
F-4	1960	8
A-7	1964	10
F-111	1970	20
F-15	1975	35
F-16	1982	45
B-2	1990	65
F-22	2000	80

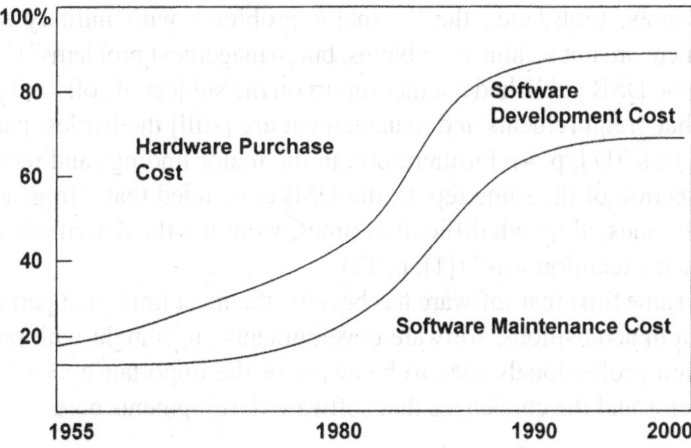

Fig. 5.1 Software and hardware costs as a percentage of total system costs since 1955 (data available online at http://software.ssu.ac.kr/SE_page/06_class1. ppt#13).

features and functions as promised. Only 16.2% of projects were successful by all measures. Of the 70% of projects that were not successful, over 52% were partial failures, in the sense that they were completed, but were over budget, over time, or did not contain all functions and features originally required. Thirty-one percent of the projects were complete failures, that is, they were abandoned or cancelled.

In 1987, the Defense Science Board (DSB) reported that, "In spite of the substantial technical development needed in requirements-setting, metrics

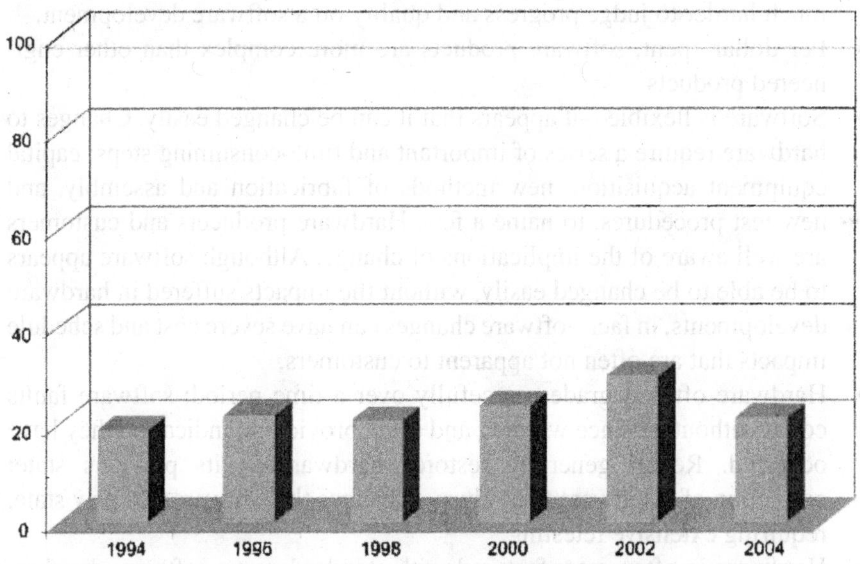

Fig. 5.2 Percentage of successful software projects for 1994–2004 [3–6].

and measures, tools, etc., the ... major problems with military software development are not technical problems, but management problems" ([8], p. 1). In 2000, the DSB published another report on the subject of software process, finding that "requirements and management are [still] the hardest part of the software task"([1], p. 4). Furthermore, in the major findings and recommendations section of the same report, the DSB concluded that, "In general, the technical issues, although difficult at times, were not the determining factor. Disciplined execution was" ([1], p. 12).

At the same time that software has become the most important part of many DoD system acquisitions, software developments are fraught with problems. Acquisition professionals need to be aware of the importance of software in new systems and the challenges that software developments pose.

5.3 Challenges of Software Project Management

The record of software project success lags that of hardware developments, indicating that managing software developments is more difficult than managing hardware developments. Some of the fundamental differences between software developments and hardware developments are as follows:

- Software is a logical, rather than a physical system. Therefore, the development cannot be approached as a manufacturing problem because software is a product that is created by peoples' minds.
- Software is invisible. It is relatively easy to measure progress of a hardware development. The number of parts produced can be counted, and quality can be checked by utilizing well-known testing techniques. It is much harder to judge progress and quality on a software development.
- Per dollar spent, software products are more complex than other engineered products.
- Software is flexible—it appears that it can be changed easily. Changes to hardware require a series of important and time-consuming steps: capital equipment acquisition, new methods of fabrication and assembly, and new test procedures, to name a few. Hardware producers and customers are well aware of the implications of change. Although software appears to be able to be changed easily, without the impacts suffered in hardware developments, in fact, software changes can have severe cost and schedule impacts that are often not apparent to customers.
- Hardware often degrades gracefully over a time period; software faults come without advance warning and often provide no indication they have occurred. Repair generally restores hardware to its previous state; correction of a software fault always changes the software to a new state, requiring extensive retesting.
- Hardware is often manufactured with standard parts; software developments have traditionally made little use of existing components. Thus,

estimating schedule and budget for software developments is inherently more error-prone than for hardware developments.

- Software developers have different skills and traits than other production workers.

The Standish Reports also determined indicators for project success and failure.

The top four indicators found in challenged projects were 1) lack of user input, 2) incomplete requirements and specifications, 3) changing requirements and specifications, and 4) lack of executive support.

The top four indicators found in failed projects were 1) incomplete requirements, 2) lack of user involvement, 3) lack of resources, and 4) unrealistic expectations.

These indicators point to requirements issues as one of the common themes in software project failures. The effect on project cost of changes to software, including changes in requirements as a function of time in the software project development, is shown on Fig. 5.3.

Past experience, as shown on Fig. 5.4, indicates that approximately 15% of a project's budget should be devoted to requirements development (in order to minimize cost overruns due to missing or erroneous requirements found in later phases of the development). Additionally, approximately 15% of the development schedule should be devoted to requirements development [8].

There are commonly occurring factors that can lead to a lack of time spent on requirements analysis and development. Schedule pressure and

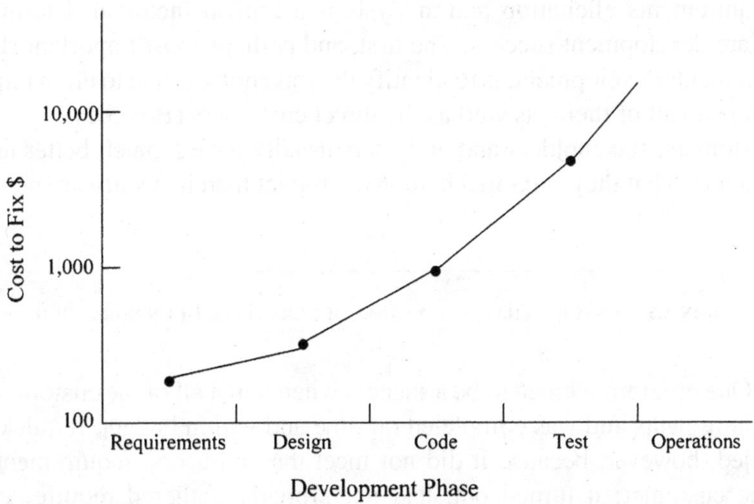

Fig. 5.3 Relative cost to fix errors or change requirements as a function of software development phase [8].

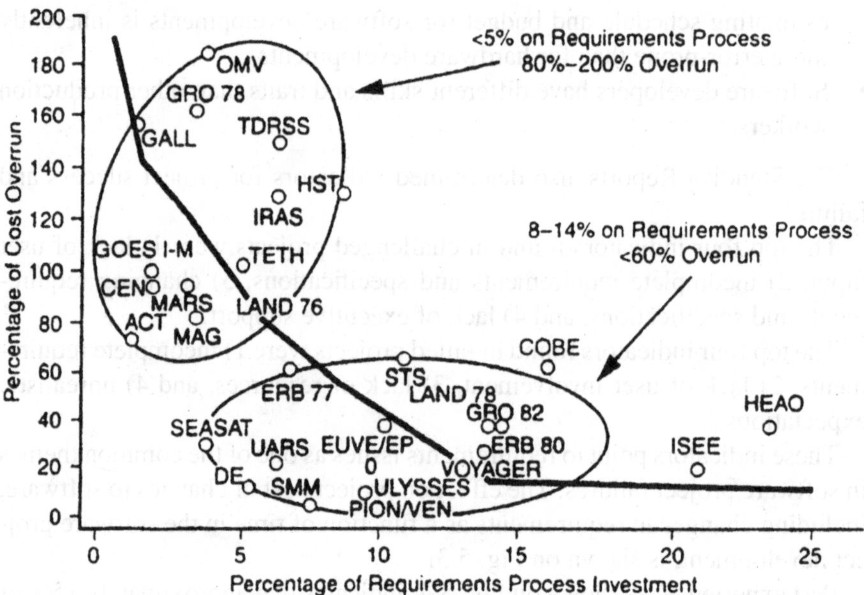

Fig. 5.4 Effect of investment in requirements development on cost overruns in software developments [9].

upper-management pressure to show tangible progress—meaning code produced—can conspire to reduce the time spent on requirements analysis. A reduction in the time spent on requirements analysis can give the illusion that schedule is being met, but eventually, there will be a heavy price paid in additional time and money spent to correct earlier short cuts.

Requirements elicitation and analysis is a critical factor in determining software development success. The first, and perhaps most important step in requirements development, is to identify the stakeholders and to elicit requirements from all of them, as well as the direct customers (Box 5.1).

Customers, stakeholders and end users usually have a much better understanding of what they want in a hardware product than in a software product.

Box 5.1 Ensure That All Stakeholders Have Been Surveyed

One program seemed to be a success when it met all of the customer's requirements and was completed on time and within budget; it quickly failed, however, because it did not meet the end-users' requirements. The customer; it turned out, had not properly gathered requirement from the end-users.

A typical customer or user statement is: "I'll know it when I see it." Some software development models, methods, and processes have features that aid in meeting the challenges of determining requirements. These will be briefly described in the following section.

Although it is essential to have customers', stakeholders', and users' inputs during requirements development, it is often difficult to get them to allocate their time for requirements elicitation. Customers, stakeholders, and users have jobs to do and usually do not recognize the importance of stating requirements correctly early in the project. Also, requirements that the customer insists are firm often turn out not to be (Box 5.2).

BOX 5.2 REQUIREMENTS SHOULD ALWAYS BE CHALLENGED AND WELL-UNDERSTOOD

A classic example of incorrect requirements occurred in the 1930s in the U.S. aircraft industry. TWA issued an RFP for a new trimotor airliner. Douglas Aircraft wanted to bid but had a new twin-engine aircraft design. Douglas Aircraft asked TWA what drove the requirement for a trimotor design. TWA said that they wanted to be able take off safely with one engine out and also wanted to be able to fly safely at 8000 feet (the highest point of TWA's transcontinental route) with one engine out. Douglas Aircraft assured TWA that those requirements could be met with a twin-engine design. Douglas Aircraft bid and won the contract. Thus was born the DC-1, which quickly evolved into the DC-3 that captured 80% of the commercial airline business in the United States by the end of the 1930s. The important lesson is that customer requirements should never be automatically assumed permanent—instead they should be challenged and clarified.

Software developments have significant differences from hardware developments and are more challenging to manage. Lack of firm, correct requirements is a recurring issue, and management must ensure that ample time and budget are devoted to requirements elicitation and analysis.

- "In architecting a new software program, all the serious mistakes are made in the first day."—*Robert J. Spinrad* [10].
- "The beginning is the most important part of the work."—*Plato* [11].

5.4 *SOFTWARE STANDARDS, CMM, AND LIFE-CYCLE MODELS*

In response to what had become known as the software crisis, the DoD sponsored the creation of the Software Engineering Institute (SEI) at the

Carnegie Mellon University in 1984. The SEI was tasked with finding ways to improve the software engineering processes. One of these improvements is the capability maturity model (CMM), which is now in use in the DoD and commercial industry. The CMM serves as a means to identify key practices to improve organizations' software development processes and proposes models to encompass systematic advancements in various aspects of the process. The SEI specialized the original model to address software (SW-CMM), management of human-resources (P-CMM), systems engineering (SE-CMM), integrated product development (IPD-CMM), and software acquisition (SA-CMM); these models, except for SA-CMM, are now part of the capability maturity model integration (CMMI). The software CMM has five levels of organization maturity, with level one being the lowest level of maturity. CMM level 3 requires, among other things, that the developer organization have a defined, documented process in place that is used throughout the organization for all software developments. An organization that uses an ad hoc method for developing software would be at CMM level 1. An organization typically must invest a minimum of a least $1 million and one or more years to implement the key practices to go from one CMM level to the next higher level. This makes obtaining audited CMM-level status problematic for small- to medium-sized software development organizations. Yet, even small organizations can, and should, institute documented, repeatable software development processes. A CMM level 3 or higher organization is highly capable and is much more likely to deliver a quality product close to planned budget and schedule than a lower-rated organization.

The most current standard governing software developments is the Institute of Electrical and Electronics Engineers/Electronic Industries Alliance (IEEE/EIA) 12207, *Standard for Information Technology—Software Lifecycle Processes* [12], a standard that establishes a common framework for software developments. This standard replaced MIL-STD-498 in 1998 [13] for the development of DoD software systems. A key feature of IEEE/EIA 12207 is the emphasis on the need for the developer to have a defined software development process that includes software project management, as well as to have an engineering process capable of developing the software. In this sense, IEEE/EIA 12207 and the CMM level 3 are well aligned.

IEEE/EIA 12207 includes several illustrative software development life-cycle models, the two most common being the waterfall model and the evolutionary model, of which the spiral model is the most often cited. The waterfall model is shown on Fig. 5.5.

Requirements analysis and definition is performed as the initial step of the waterfall model, followed by software design, coding, and testing and integration. The waterfall model has the advantages of being easier to manage than evolutionary developments, can be scheduled to correspond to completion of the specific individual phases of the project, and often is the fastest and most

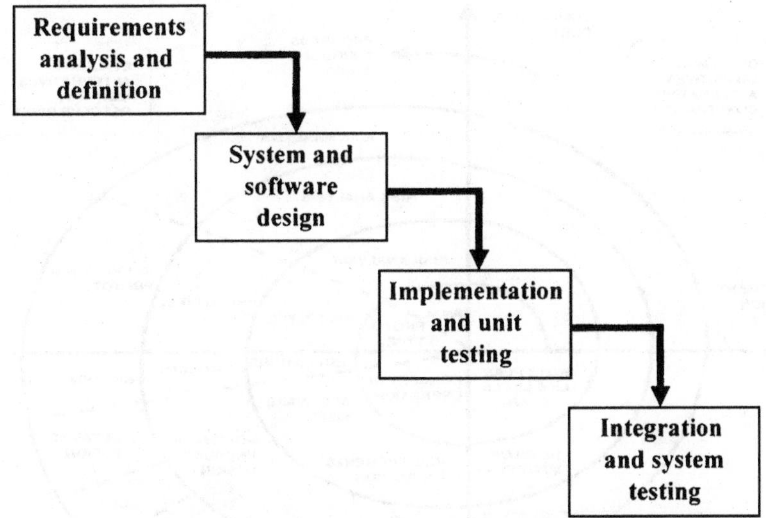

Fig. 5.5 Waterfall model for software development.

economical method of developing software—provided the requirements are well known. If there are requirements changes, the process returns to the initial phase, usually at considerable cost and schedule delay.

One alternative approach is prototyping, as shown on Fig. 5.6. In this approach, the developer elicits an initial set of requirements and builds an initial prototype. The customer tries the prototype and provides additional requirements definition to the developer. The process shown in Fig. 5.6 cycles until the customer is satisfied with the prototype. The final prototype should not be thought of as a deliverable system, but rather as an executable set of requirements from which the deliverable system is developed using a waterfall approach.

The spiral model of development [14], as illustrated in Fig. 5.7, combines the features of the prototyping model and the waterfall model. The spiral

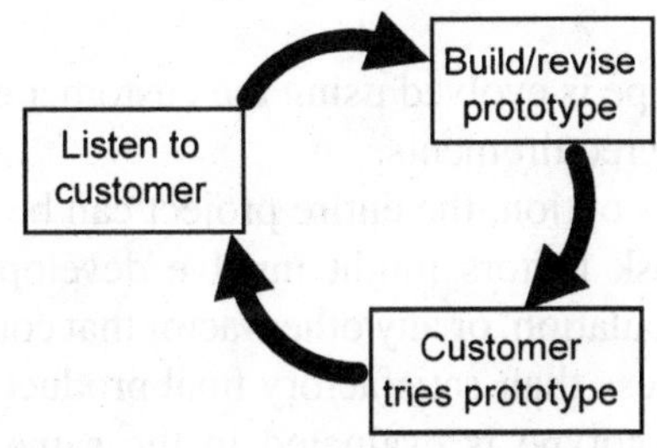

Fig. 5.6 Prototyping.

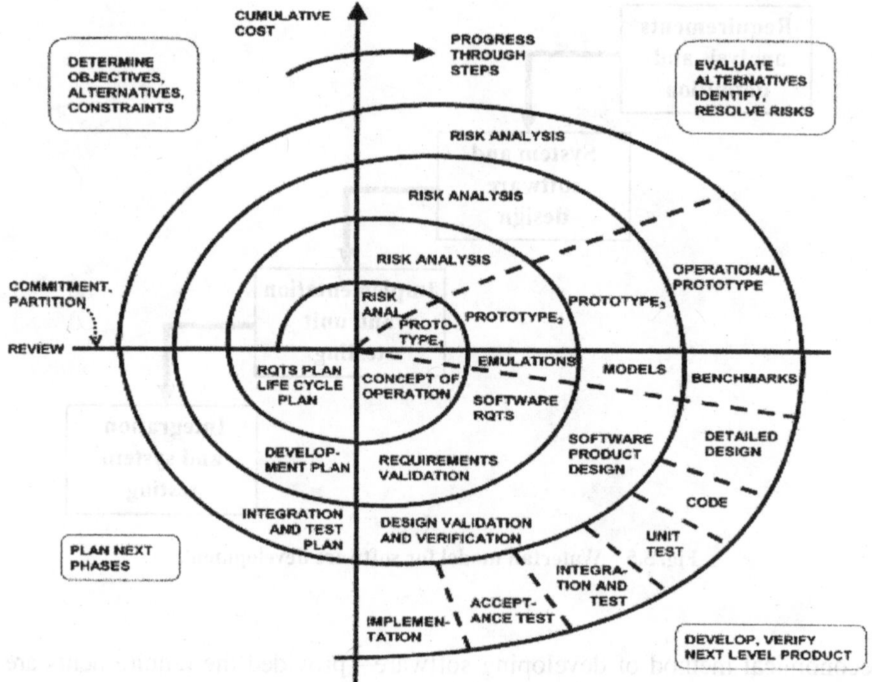

Fig. 5.7 Spiral model for software development [15].

model is intended for large, expensive, and complicated projects, often those that have considerable requirements risk. Each phase begins with a design goal and ends with the customer (who might be internal) reviewing the progress thus far. Analysis and engineering efforts are applied at each phase of the project, with an eye toward the end goal.

The steps in the spiral model can be generalized as follows:

1) The initial requirements are defined in as much detail as possible.

2) A preliminary design is created for the new system.

3) A first prototype of the new system is constructed from the preliminary design. The customer evlauates the prototype in terms of its strengths and weaknesses.

4) A second prototype is evolved using the customer evaluation of the first prototype to refine the requirements.

5) At the customer's option, the entire project can be aborted if the risk is deemed too great. Risk factors might involve development cost overruns, operating-cost miscalculation, or any other factor that could, in the customer's judgment, result in a less-than-satisfactory final product.

6) The existing prototype is evaluated in the same manner as was the previous prototype, and, if necessary, another prototype is developed from it.

7) The preceding steps are iterated until the customer is satisfied that the refined prototype represents the final product desired.

8) The final system is constructed, based on the refined prototype, using a waterfall approach.

The spiral model differs from the prototyping model in that there is considerable emphasis on risk analysis at the start of each of the spirals and subsequent resolution of risks during the remainder of each of the spirals. The risk-analysis phase is the most important part of the spiral model. In this phase, all possible (and available) alternatives that can help in developing a cost-effective project are analyzed. This phase has been added specially in order to identify and resolve all of the possible risks in the project's development. If risks indicate any kind of uncertainty in requirements, prototyping can be used to find out possible solutions to deal with the potential changes in the requirements.

Spiral development has the primary advantage of expecting and responding to uncertainties in requirements and is more able to cope with the changes inherent to software development. Disadvantages of the spiral approach are that it is harder to manage than the waterfall approach, more difficult to estimate total development budget and schedule at the start, and requires highly skilled people in the area of planning, risk analysis and mitigation, development, customer relations, etc. (Box 5.3).

The DoD has recently rewritten the defense acquisition regulations to incorporate evolutionary acquisition, an acquisition strategy designed to mesh well with spiral development. The choice of development approach, however, should be driven by the assessment of requirements. If requirements are well-known, a waterfall method is appropriate; if not, a spiral method, or other evolutionary method, is recommended.

Software systems can be decomposed into CSCIs (computer software configuration items). CSCIs are defined to be an aggregation of software that satisfies an end-use function, can be managed separately, and has separate configuration management. CSCIs can also be developed using separate development approaches.

There is often confusion between the spiral model and incremental development—in which software is developed as a sequential set of builds. A *build* is a version of software that meets a specified subset of requirements. The term also can refer to the period of time during which such a version is developed. Incremental development is a strategy that can be applied to the waterfall model, the spiral model, and other development models. When an incremental strategy is applied to the spiral model, each loop through the spiral results in an increment that is delivered to the customer.

Developers of commercial software systems have reported great success using what are known as agile processes, an example of which is extreme programming. Extreme programming (known as XP) projects unanimously

Box 5.3 Term "Spiral Model" Is Often Misapplied

The term "spiral model" is often misapplied. A three-star flag officer, at the kickoff meeting for the development of a new joint services system, announced that the system would be developed using a spiral method and that it would be completed in two years and two spirals. How, before the development started, did he know that it would take two years and two spirals?

report greater programmer productivity than other projects within the same corporate environment. Customers are required to be active participants in XP developments. Customers write user stories, in the format of about three sentences, as things that the system needs to do for them. Then, the developer must create one or more automated acceptance tests to verify that the user story has been correctly implemented. When the time comes to implement the story, developers will go to the customer and receive a detailed description of the requirements face to face, but the customer is relieved of the necessity of writing a detailed requirements document. Developers estimate how long the stories might take to implement, and the customer negotiates a selection of user stories to be included in each scheduled release. This process helps ensure that the scope of the development is consistent with budget and schedule, resulting in a high probability of project success. Code is created by two people working together at a single computer. This *pair programming* increases software quality without impacting the time to develop it. Code integration is accomplished by each paired team releasing their tested code, in turn, to a controlled code repository. Developers should be integrating and releasing code into the repository every few hours—whenever possible. This almost continuous integration avoids or detects compatibility problems early. XP has been primarily used on small projects involving 2 to 10 developers. That fact, and the requirement of the customer as full-time member of the development team, makes XP difficult to implement on most DoD software developments.

There are specific software development methods, in addition to development models, that also aid requirements management. UML, an object-oriented modeling language, for example, utilizes use cases to help requirements elicitation and analysis. *Use cases* are typical interactions that a user has with the system in order to achieve some goal. Use cases provide the basis of communication between customers and developers. The developer should discover most use cases during requirements analysis, particularly the most important ones, but the system does not have to be constructed in a waterfall manner when using UML. If the key use cases are correctly identified, the system architecture will support future extensions.

A concern in software developments is that the cost to maintain software can be the dominant part of the total software life-cycle cost. These maintenance cost drivers are established during the initial software development. For example, program managers can often decrease maintenance costs by urging software developers to use object-oriented methods. But, at the same time, object-oriented methods can add to the development cost and can result in compiled code that is more complex, larger, and that requires more processing power and memory to execute.

No one development approach is best for all projects. The approach should be tailored to project size and requirements uncertainty, among other factors.

5.5 ESTIMATION AND PLANNING

Possible explanations for poor performance on software projects include poor project management and requirements problems. Another explanation is that budgets and schedules are inaccurately and overoptimistically estimated. Cost estimation of software projects is certainly more difficult than for hardware projects because, as stated earlier, software is a logical, rather than a physical system. Software is a product of peoples' creative process and is much less predictable than a manufacturing problem.

A software cost estimate can be based on algorithmic models that relate software development cost either 1) to software size estimates in terms of lines of code or 2) to software functionality estimates. Often, these algorithmic models also include additional estimation factors related to elements such as the complexity of the development, the ability of the development team, and development constraints. One of the most well-known algorithmic models based on software size is COCOMO, developed by Barry Boehm [16]. This was later extended to COCOMO II [17] to account for object oriented developments and also evolutionary development models.

The best algorithmic models generally, at best, provide an estimate that is no closer than ±20% of an actual cost. To use algorithmic models, the models have to be calibrated to the developer organization. This requires that the developer organization have a well-adjusted database and that it has standardized its development methods—another reason for the importance of a developer having instituted a documented, repeatable software development process.

Cost estimates can also be based on tasks or activities tied to work breakdown structures. The best practice when estimating software developments is to use multiple estimation techniques, including algorithmic models, activity-based estimates, top-down expert opinions and costing by analogy to confirm individual estimates, or to understand differences in estimates and gain insight into a more accurate estimate.

COCOMO also provides a means of estimating the project duration, given an estimate of the software size and complexity. Standard scheduling techniques also can be applied, starting with the development of a work breakdown structure. However, these techniques cannot be readily applied when using a spiral development approach; in this case, the schedule can usually only be developed in a phased manner.

An essential part of good planning is a detailed risk-management plan. There are numerous sources of software development risk in addition to the obvious risks of cost and schedule overruns. The SEI has a detailed software project risk identification checklist that is an excellent starting point for the development of a sound risk-management plan.

Personnel risks are especially important to consider in software developments because software developers pose special challenges and the attrition problem in the software industry can be disastrous for a midstream software project. Brooks' famous law [18], "Adding manpower to a late software project makes it later," explains why it can be a disaster.

Software projects, like other projects, require careful estimation and planning. Yet, software projects have the additional challenge of frequently inaccurate cost estimation and schedule estimation. Software is the product of a creative process, and therefore it is less able to be approached as a known, well-understood process.

5.6 MONITORING, METRICS, AND QUALITY ASSURANCE

It is difficult to judge progress in software development because software is invisible. Thus, monitoring progress of software development requires the use of metrics and other indicators that can be related to progress.

IEEE/EIA Standard 12207 includes examples of metrics that can be used to judge the progress of software development and, in some cases, predict future performance. With metrics, monitoring and project control will be based on facts. Example metrics include number of project staff, effort (by work package), labor cost, elapsed time to completion, newly developed code, number of defects, and number of requirements changes. The most important metrics are those that will give advanced indication of potential future cost and schedule problems. For example, referring back to Fig. 5.3, if the metric "number of requirements changes with time" shows requirements volatility after the initial requirements-analysis phase, the software project will almost certainly experience cost overrun and schedule slippage. Likewise, if size of the code is used to estimate development cost, then periodic updates on the size estimate as a measure of potential future cost growth would be crucial. Metrics that are required by the customer for monitoring purposes need to be specified in the contract and gathered, usually on a weekly basis, by the developer. Customer oversight is required to ensure that the metrics are timely and meaningful and not treated by the developer as a "check-the-box" activity.

Monitoring earned value is also recommended; careful attention should be paid to the quality of products marking the completion of work packages. A unit development folder (UDF) is an additional means of monitoring progress. A UDF is a formalized programmer's notebook, is used to overcome the 90% completion syndrome, and is given to a programmer-analyst who is the developer of a unit of code and is the UDF custodian from detailed design through integration.

Software quality must be incorporated throughout the development process. It is not possible to go back and add quality, nor to test in quality. Internal and external reviews are both important tools that help achieve high quality as well as measures of progress. Examples of internal reviews are software walkthroughs and software inspections. Walk-throughs occur when a software developer asks peers to review and critique a product in development. Walk-throughs are informal, do not require much overhead, and are designed to catch flaws early in the process. Software inspections are more formal in that trained teams review software products using checklists based on the development organization's past experience. Good software inspections can be more than 95% effective in finding flaws, and they should be a key element of a well-defined software development process.

External reviews are typically held at the conclusion of a major phase of a software development, such as the conclusion of requirements development and analysis and conclusion of preliminary design or conclusion of detailed design, all of which align well with a waterfall approach. External reviews involve the customer, users, and development team and provide customer management visibility into project progress. At the conclusion of an external review, the customer management formally decides (in written notification to the developer) to 1) allow the project to continue on to the next phase—providing that issues raised during the review are fixed in accordance with the schedule provided in the review minutes, 2) require that the review be redone at a specified future date, or 3) cancel the project because of lack of progress or excessive risk. The usual outcome of external reviews is approval by the customer to continue to the next phase. The customer and developer must work collaboratively well ahead of the review in order to have high assurance of a successful outcome. This collaboration is best achieved by close interaction between the developer and customer technical workers and by early transmittal of review material by the developer to the customer (Box 5.4).

Measuring progress on software developments is difficult. A variety of techniques should be used, to include internal and external reviews, and careful use of a range of metrics.

BOX 5.4 IMPORTANCE OF HONEST REVIEWS

An Army program passed its requirements review and preliminary and critical design reviews with flying colors—even though many requirements were still undetermined—because the customer and developer were focused on keeping the program on schedule. Later, in order to correct missing and incorrect requirements, the schedule slipped by two years.

5.7 CONFIGURATION MANAGEMENT, TESTING, AND MAINTENANCE

Configuration management is included in the scope of IEEE/EIA 12207. Good configuration management is important because it adds discipline to requirements change management through inclusion of a change control board (CCB) function and ensures necessary control over project baseline items. One of the configuration management challenges in spiral development is to manage prototype configurations, models and simulation results, and document and manage customer feedback. Configuration management is also needed in order to provide proper version control of released software (Box 5.5).

Testing is sometimes considered the most important phase of software development because it is the last opportunity to catch errors before the software is delivered for operational use. Planning for testing must begin in parallel with requirements analysis. Every requirement must be testable, and, as each requirement is written, a corresponding test approach must be

BOX 5.5 IMPORTANCE OF VERSION CONTROL

SPAWAR encountered problems with software that had been delivered by an external supplier. The supplier manager was shocked and surprised that SPAWAR reported problems using the software are because the software was working perfectly at the developer site. On further investigation, it was found that the software SPAWAR was attempting to use was a version that had not gone through integration testing and had been sent to SPAWAR by individual developers who were trying to be responsive to what they determined was a customer need for a quick release. There was no configuration management or version control at the developer site.

identified and incorporated into the test plan. This is accomplished explicitly in extreme programming because an acceptance test must be developed for every user story. Schedule often has a major impact on testing. Because testing is the last phase of software developments and because schedules of prior phases are often overrun, there can be pressure to reduce the amount of test coverage in order to maintain schedule. The decision to cut back on testing almost always results in flaws that are uncovered during operational use and are very costly and time consuming to fix. A better option is to maintain planned test coverage and extend the schedule.

Highly successful software will be in operational use for an extended time, during which the software must be maintained. There are three categories of software maintenance: *corrective*, changes to correct previously undiscovered errors, found to account for 17% of all maintenance on the average; *adaptive*, changes required because of changes in the environment of the software, accounting for 18% of all maintenance; and *perfective*, changes that improve the system, representing the remaining 65% of all maintenance. Often, most of the life-cycle cost occurs after software delivery during the maintenance of the software, yet the drivers to reduce maintenance cost are built in during the software development. Factors that tend to reduce maintenance cost are effective modularity of software design (best obtained by use of object-oriented methods) and good documentation and configuration management. Typically, the development program manager is rewarded based on his/her completion of a program within budget and schedule. Thus, anything that would enhance downstream maintenance, but would increase development time and cost, is likely to be omitted. To counter this tendency, acquisition professionals must build incentives into the development that reward the program manager for reducing future maintenance costs (Box 5.6).

Configuration management stabilizes software developments and ensures that developers and customers have a common understanding of baselines. Testing must be planned and scheduled for the initial stages of requirements

Box 5.6 Plan for Expansion

A software system that met customer requirements perfectly was delivered. Other potential customers and users saw the software in use and wanted it adapted to their needs. However, the software had not been designed for easy modification, and it quickly came to an end of useful life.

analysis, and the urge to cut back on testing to maintain schedule should be strongly resisted. Factors to reduce long-term maintenance costs need to be encouraged during the development phase.

5.8 Commercial Off-the-Shelf Software and Software Reuse

Recent DoD regulations encourage the use of commercial off-the-shelf (COTS) software in lieu of custom software developments. Use of COTS presents both an opportunity and a risk. The opportunity is that time, money, and risk can be reduced if appropriate COTS solutions are available. The risk is that selecting the wrong COTS software component might be more expensive than fixing problems in custom-built software. To best utilize COTS solutions, current DoD policy requires system-acquisition agents to avoid government-unique requirements, avoid restrictive statements of need and avoid detailed specifications. Current literature indicates that an integrated COTS software solution works best in an environment of flexible requirements.

Releases of COTS software components can arrive regularly and are difficult to reintegrate; the developer has no control over product evolution and must compensate by staying aware of pending changes. These and other issues require that a proactive, disciplined, and systematic COTS software risk management approach be in place early in the development.

COTS risks are in the areas of technology, vendor, and product. Examples of technology risks are technologies that will not last, that will undergo significant change, or that have poor product competition. Examples of vendor risks are buying from a vendor that will not last, has a limited product line, is unable to maintain a product line because of poor development/maintenance practices, or is unable to adapt a product to a new environment/technology. Depending on the government's market share of a particular piece of COTS software, the government might no longer be able to affect desired requirements or constraints on the COTS component; conversely, another customer might be able to bend the product to satisfy their needs.

Maintenance turnaround time by the vendors can be a significant problem. Maintenance on identified problems is provided by the COTS software vendor, but problem investigation and identification by the integrator are the most costly parts of COTS software maintenance.

Examples of product risks are buying a product that will not last, that will undergo significant change or will require a significant number of upgrades/patches, that will be difficult to integrate and will require additional documentation, that will compromise system security or personnel safety, or that has expensive maintenance fees. SPAWAR has used a COTS software risk checklist that addresses detailed risk issues in the areas of technology, vendor, and product as a means to assess risks in using COTS products [19].

Software reuse can offer cost savings, reduce development time, reduce risk, and increase reliability. *Reuse* is the process of implementing new software systems from preexisting software and reapplying knowledge about one system to another similar system in order to reduce development effort.

In spite of its compelling advantages, there has been only limited success in reuse of software. A key requirement for software reuse is *domain analysis*—a process by which information used in developing software systems is identified, captured, and organized with the purpose of making it reusable in new systems. Domain analysis is typically a slow, unstructured learning process; it is a phase in the software life cycle in which a domain model (describing common functions, data, and relationships of subsystems in the domain), a dictionary (defining terminologies), and a software architecture (describing packaging, control, and interfaces) are produced. Application areas in which software reuse has been most successful are those where there has been successful domain analysis (Box 5.7).

BOX 5.7 SUCCESS IN SOFTWARE REUSE

International Submarine Engineering, Limited, of Port Coquitlam, BC, Canada, has achieved outstanding success reusing software in their product lines of remotely operated and autonomous underwater vehicles, unmanned surface water vehicles, manned submarines, and robotic manipulators. This success is caused by outstanding domain knowledge gained over more than 30 years of company experience and very modular product line architectures.

The best approach when attempting to maximize the reuse of software is to first search for available reusable components and then adapt the requirements and the software architecture to best utilize the components. Much like incorporation of COTS software, the reuse of software is enhanced by keeping requirements flexible.

Challenges in reusing software are to find appropriate reusable components, to understand the components, and to have confidence in their suitability. This requires good design documentation, commented code, and well-documented test data. The components must include associated information describing how they can be reused.

COTS software and reused software both offer advantages of reduced cost, development time, risk, and increased reliability. At the same time, COTS software and reused software both pose substantial risks if incorporated inappropriately. Management must carefully assess the risks before committing to COTS and software reuse.

5.9 CONCLUSION

Software developments are becoming increasingly important and pose special challenges that require close management attention and understanding. Good requirements development and analysis can make or break an acquisition. Thus, this is usually the single most important step in a software development. A variety of management techniques will help ensure software development success.

5.10 ADDITIONAL RESOURCES

The following organizations and Web sites are provided for further reference:

- Software Engineering Institute: www.sei.cmu.edu
- Software Productivity Consortium: www.software.org
- Software Technology Support Center: www.stsc.hill.af.mil (Also the home of *Crosstalk* magazine)
- USC Center for Software Engineering: http://sunset.usc.edu/cse/

REFERENCES

[1] Defense Science Board, "Report of the Defense Science Board Task Force on Defense Software," Office of the Under Secretary for Acquisition and Technology, Washington, D.C., Nov. 2000.

[2] U.S. General Accounting Office, "Contracting for Computer Software Development—Serious Problems Require Management Attention to Avoid Wasting Additional Millions," FGMSD-80-4, Washington, D.C., 9 Nov. 1979.

[3] SPAWAR Systems Center, *Software Project Management Course Notes*, Software Engineering Process Office, San Diego, CA, 1999, p. 11.

[4] Standish Group, "CHAOS: A Recipe for Success, 1998," Standish Group International, West Yarmouth, MA, 1999, pp. 2–5.

[5] Standish Group, "CHAOS Reports," Standish Group International, West Yarmouth, MA, www.standishgroup.com [retrieved July 2007].

[6] Standish Group, "The Standish Group Report—CHAOS 1994," Standish Group International, West Yarmouth, MA, 1995, pp. 2–6.

[7] Standish Group, "What Are Your Requirements?," Standish Group International, West Yarmouth, MA, 2003, p. 1.

[8] Defense Science Board, "Report of the Defense Science Board Task Force on Military Software," Office of the Under Secretary for Acquisition, Washington, D.C., Sept. 1987.

[9] SEMATECH Technology Transfer #92111389A-TRG, 5 March, 1993.

[10] Spinrad, Robert J., Lecture, *The Art of Systems Architecting*, 2nd ed., edited by M. Meir and E. Rochtin, CRC Press, New York, 2000, p. 30.

[11] Plato, *The Republic*, translated by F.C. MacDonald, Oxford Univ. Press, New York, 1945.

[12] Inst. of Electrical and Electronics Engineers/Electronic Industries Alliance, IEEE/EIA 12207, *Standard for Information Technology—Software Lifecycle Processes*, New York, May 1998.

[13] Dept. of Defense, *Software Development and Documentation*, MIL-STD-498, Washington, D.C., Nov. 1994.

[14] Hooks, I., *Managing Requirements*, Compliance Automation, Inc., 1994, pp. 1, 2; http://www.complianceautomation.com/.

[15] Boehm, B.W., "A Spiral Model of Software Development and Enhancement," USC-CSE-1988-500, USC-Center for Software Engineering, Los Angeles, May 1988.

[16] Boehm, B.W., *Software Engineering Economics*, Prentice–Hall, Upper Saddle River, NJ, 1981, pp. 57–156, 344–531.

[17] Boehm, B., Clark, B., and Devnani-Chulani, S., "Calibration Results of COCOMO II.1997," USC-CSE-97-507, USC-Center for Software Engineering, Los Angeles, 1997.

[18] Brooks, F. P., *The Mythical Man-Month*, Addison Wesley Longman, Reading, MA, 1982, pp. 14–26.

[19] Hensley, B. J., "Development of a Software Evolution Process for Military Systems Composed of Integrated Off-the-Shelf (COTS) Components," Master's Thesis, Dept. of Computer Science, Naval Postgraduate School, Monterey, CA, March 2000.

STUDY QUESTIONS

5.1 Why are software developments more difficult to manage than hardware developments?

5.2 Why is requirements analysis so important in software developments?

5.3 What are key differences in various software development models?

5.4 When is using a spiral model a preferred approach?

5.5 Does using COTS software always reduce development cost and risk?

Chapter 6

LOGISTICS AND SUSTAINMENT

BRAD NAEGLE*
NAVAL POSTGRADUATE SCHOOL, MONTEREY, CALIFORNIA

LEARNING OBJECTIVES

- Understand that logistics is a system performance attribute
- Describe acquisition's role in logistics supportability performance
- Understand logistics supportability design impact on total ownership cost
- Describe the 10 logistics elements
- Describe design for supportability performance in the systems engineering process
- Describe supportability analyses
- Understand system failure or wear-out patterns
- Understand the importance of driving supportability performance in the request for proposal
- Describe supply-chain management
- Describe contractor logistics support
- Describe performance-based logistics (PBL)

6.1 INTRODUCTION: LOGISTICS AS PERFORMANCE

A system's logistics supportability is a major component of its performance and should be treated equal to other critical performance parameters. A race car is designed for ease and speed of servicing during a pit stop because that maintainability design is a significantly important performance characteristic for winning the race—as important as the speed, cornering ability, braking, or any other aspect of the race car system. Likewise, the logistics supportability design of a weapon system impacts its combat effectiveness. Systems that are highly reliable and that can be quickly and easily returned to mission-capable

*Senior Lecturer, Graduate School of Business and Public Policy.

status when maintenance is required increase the warfighter's capability. The bottom line is that *improved logistics supportability design equals combat power.*

Software-intensive systems are becoming ever more dominant in newly developed systems, placing even more emphasis on system supportability performance. In addition to providing a high level of system functionality, the dynamic and flexible nature of software means that it will be subject to a significant number of maintenance actions—correcting defects, upgrading performance, adapting to interoperability, closing safety and security vulnerability gaps, etc. If system software is not designed for effective supportability, the system performance will likely degrade very quickly, as software maintenance will be required immediately and often throughout the systems' life cycle. Maybe more than any other element, software supportability is a key enabler of system performance and cannot be considered separately.

6.1.1 ACQUISITION'S ROLE IN LOGISTICS PERFORMANCE

The majority of a system's total ownership cost (TOC) is in the postdeployment sustainment phase and typically represents 60 to 80% of the TOC. This ratio holds true with software-intensive, net-centric warfare systems, as software "maintenance" during the sustainment phase continues to dominate TOC for those systems. Figure 6.1 depicts a typical system's cost distribution, illustrating that development and procurement costs represent the smaller portion of TOC.

It is evident that decisions affecting a system's sustainment phase have the most impact on TOC performance. The main driver of sustainment-phase cost

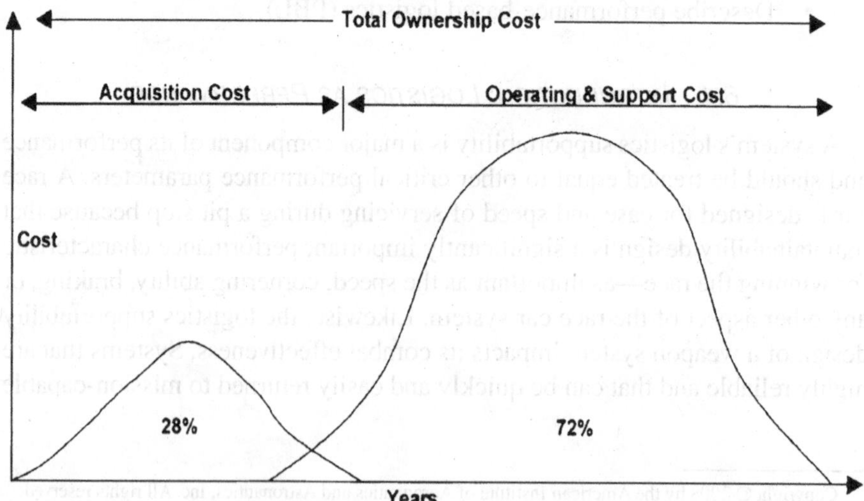

Fig. 6.1 Nominal life-cycle cost of typical DoD acquisition program ([1], p. 14).

is the system's logistics supportability performance, which can only be substantially influenced during early development prior to the production RFP.

Adding emphasis to system supportability performance, the Chairman of the Joint Chiefs of Staff Instruction 3170.01F [2] specifies that all programs must include a materiel readiness key performance parameter (KPP) in the user requirements documents. As with all KPPs, this mandatory KPP means that the threshold performance specified for a system's materiel readiness must be achieved and cannot be a consideration during tradeoff analyses.

The integration of logistics supportability performance considerations into the systems engineering process (SEP) analysis is critical in making decisions and driving the design to maximize the warfighter capability. These analyses, conducted early in the acquisition process, are fundamental in formulating the system's supportability concept—level of repair analysis, organization, personnel structure and skill-set requirements, support and diagnostic equipment, etc. In aggregate, this process provides an estimate of the system's "logistics footprint" and the "logistics burden" likely to be imposed by the deployment of the weapon system.

Systems analysis is needed for maximizing system performance, including systems supportability performance. It is easy to become focused on the end item, what "it" can do and how much "it" costs to procure, without considering the broader weapon system. When one refers to an F-16 Falcon, one might well picture Fig. 6.2.

Although this is certainly an F-16 Falcon, it certainly is not an F-16 Falcon weapon system, which is more accurately depicted in Fig. 6.3.

Fig. 6.2 F-16 Falcon.

Fig. 6.3 F-16 Falcon with support personnel and equipment.

This depiction shows support personnel, support equipment, test-measurement-diagnostic equipment (TMDE), fuel, and facilities necessary to support the mission-capable status of the fighter—elements that are integral parts of the F-16 Falcon *weapon system.* Of course, there are other elements not depicted but that are just as critical to the weapon system, including training, personnel support systems, software, supply and maintenance management systems, automation, industrial base contractors, and many, many others that must be included in *systems thinking.* The optimization of this weapon system is the key to providing combat power to the warfighter.

Logistics supportability performance becomes most evident after the system has transitioned to the deployment and sustainment phase of its life cycle, but the supportability performance *achieved* is significantly determined early in the acquisition development process. The graph in Fig. 6.4 depicts this concept.

Consistent with the depiction in Fig. 6.1, the bulk of the funding expended on a typical system occurs after deployment—in the sustainment phase. The solid black line above shows how much of the cost, reflecting the logistics supportability design, is determined extremely early in the development cycle. Approximately 80 to 90% of the supportability performance, therefore, TOC is locked in by the time the requirements have been set and communicated to potential contractors. Clearly, logistics supportability performance, like any other performance attribute, must be considered from the concept refinement phase and detailed in the technology development phase for inclusion in the system performance specification and RFP.

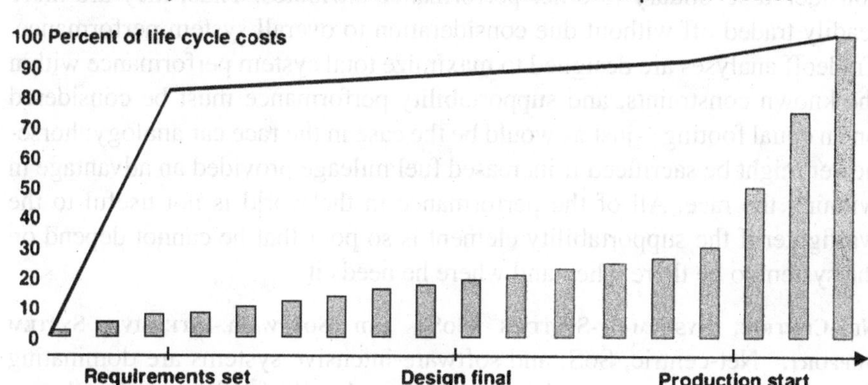

Fig. 6.4 Acquisition decisions' impact on TOC ([1], p. 15).

6.1.2 LOGISTICS DESIGN LINKAGE TO TOTAL OWNERSHIP COST

COST-AS-AN-INDEPENDENT-VARIABLE (CAIV) ANALYSIS. Too often, the wrong "cost" category is used when performing CAIV techniques supporting tradeoff analyses. The concept behind CAIV centers on tradeoff analyses based on TOC, not procurement or contract cost. For example, eliminating a built-in-test (BIT) system is likely to reduce the procurement cost, but the impact on TOC—problem diagnosis time, technician training requirements, need for TMDE, improper part/subcomponent replacement caused by faulty diagnosis, system downtime, etc.—makes that type of decision extremely costly in TOC terms.

System logistics supportability is a major component of TOC, and so tradeoff decisions involving supportability performance typically have significant impact on system TOC. The best-value decision using CAIV analysis depends on TOC being used as the "cost" category analyzed for tradeoff. Reducing system TOC through effective CAIV analyses frees scarce funding resources for other critical needs, further supporting the warfighter.

TRADEOFF ANALYSES. In addition to cost, other tradeoff analyses involving supportability performance are inevitable. The user or combat developer helps to shape the tradeoff analyses through the Joint Capabilities Integration and Development System (JCIDS) process, identifying KPPs and critical operational issues (COIs). System design features supporting KPPs and COIs are typically excluded from tradeoff consideration, as they describe critical aspects of the warfighter capabilities needed.

Supportability performance attributes are rarely designated as a system KPP or COI; therefore, most supportability elements are in the trade zone during tradeoff analyses. Worse yet, supportability elements have often been considered secondary to other performance attributes. Thus, they are more readily traded off without due consideration to overall system performance. Tradeoff analyses are designed to maximize total system performance within the known constraints, and supportability performance must be considered on an equal footing—just as would be the case in the race car analogy: horsepower might be sacrificed if increased fuel mileage provided an advantage in winning the race. All of the performance in the world is not useful to the warfighter if the supportability element is so poor that he cannot depend on the system to be there when and where he needs it.

Net-Centric, System-of-Systems (SoS), and Software-Intensive System Support. Net-centric, SoS, and software-intensive systems are dominating the methodical transformation of DoD and will fundamentally change supportability concepts, planning, and analysis. Software provides an ever-increasing portion of new systems' functionality, and software supportability performance is essential in providing the warfighter these capabilities when needed and with long-term reliability.

One common element of net-centric, SoS, and software-intensive systems is that software (most often, lots of software) is typically the significant enabler for leveraging the advantages these types of systems offer. Although hardware logistics supportability will always be important, software supportability issues are likely to surpass them in terms of volume and cost, especially as ever-larger software-intensive systems are envisioned and deployed.

The system software tends to lose its effectiveness faster than any other system component, and so software supportability is a critical element in system performance. The need for software maintenance, upgradeability, interoperability, reliability, safety, and security actions begins immediately after system deployment, occurs at a much higher rate than for hardware, and continues throughout a system's life cycle. The software design's ability to accommodate supportability actions has a major impact on system performance and TOC.

Software supportability has some aspects in common with hardware supportability and others that are in stark contrast. Supportability analyses are typically common for both software and hardware elements, as both are focused on system supportability performance. Software and hardware supportability performance is significantly determined by the architectural design, and so initial supportability concept development and planning activities for both will likely be very similar. For software life-cycle supportability planning and documentation, the computer resources life cycle management plan (CRLCMP) is typically the source document and is continually refined throughout the development and operational support phases of a system. In concert with the CRLCMP, the computer resources integrated support document (CRISD) provides information needed to plan for the life-cycle support of software-intensive systems. The CRISD also includes the software developer's plans for transitioning software life-cycle support to the government or another software sustainment entity.

Influencing the system design for improved supportability performance is one area in which software differs significantly from hardware. Hardware-oriented engineering environments tend to be much more mature than the software engineering environment, and that lack of maturity must be addressed in the requirements analysis/performance specification development/request for proposal (RFP) process. For example, the mature automotive industry will include design provisions for replacing oil, filters, tires, batteries, and other normal wear-out items, as those are widely accepted industry norms; indeed, any competent manufacturer will provide them automatically. There are no corresponding industry-wide standards for software supportability design. Without this engineering maturity, the software engineer designs the architecture in response to the performance specification and other explicit requirements provided. Software supportability performance needed by the warfighter must be provided in significant detail to drive the software design to meet the required capability.

The cost of "maintainers" is another area in which software is different than hardware. Hardware maintenance personnel are typically technicians with skill sets focused on the maintainability aspects of a system or family of systems. Software maintenance personnel are usually software engineers and are focused on all of the software supportability aspects—including maintenance, upgradeability, interoperability, and reliability. They also have significant responsibilities in software safety and security. Most often, software support will not be organic, but provided through some form of contractor logistics support (CLS) arrangement. The cost of the necessary software engineers for system support will likely be more expensive than comparable hardware maintenance technicians.

6.1.3 Logistics Elements

Planning for logistics supportability can be more comprehensive by considering the 10 logistics elements: *maintenance planning; manpower and personnel; supply support; support equipment; technical data; training and training support; computer resources support; facilities; packaging, handling, storage, and transportation; and design interface.* This analysis must be conducted at every level of maintenance envisioned for the system to help capture the total logistics burden and more accurately estimate TOC. The logistics-elements analysis technique facilitates an understanding of the systems view by including elements that define the derived and implied supportability requirements, especially by capturing those that are not apparent through the system's acquisition process.

MAINTENANCE PLANNING. In sequencing the analysis of the 10 logistics elements, performing maintenance planning analysis first makes sense. As just stated, the 10 logistics element analysis must be performed at every level of maintenance, and determining a system's number of maintenance levels is one of the key products of performing maintenance planning. It constitutes a sustaining level of activity, commencing with the development of the maintenance concept and continuing through the accomplishment of supportability analyses. Effective maintenance planning conducted in the concept refinement and technology development phases helps to drive the system design for maintainability and to define reliability requirements.

MANPOWER AND PERSONNEL. Manpower and personnel are major drivers of TOC and significant considerations in tradeoff analyses at every level. *Manpower* (the total number of people needed to effectively operate and maintain the system) and *personnel* (the level of education and training and skill sets and aptitudes required to perform their functions within the system) are critical effectiveness and suitability drivers of the system under development. As one of the most expensive components in any system, design decisions impacting the number of personnel or their necessary education, training, and skill levels have a major impact on TOC performance.

SUPPLY SUPPORT. A system's supply-support requirements are a major element in its logistics burden and "logistics footprint." Supply Support includes spares (repairable units, assemblies, modules, etc.), repair parts (nonrepairable components), consumables (munitions, lubricants, filters, coolant, etc.), and related inventories (welding supplies, cleaning solutions, TMDE consumables, etc.) necessary to sustain system operation over extended periods of time. Supplies required for initial provisioning, procurement activities, and documentation are also included in this category. Supply support can be linked to a system's operational availability A_o in cases where numerous, but not time-consuming, maintenance actions are required. Such a system would require significant supply support to maintain a high A_o.

SUPPORT EQUIPMENT. Support equipment is another element that can have a significant impact on a system's logistical footprint. The F-16 support equipment depicted in Fig. 6.3 is a good example of the amount of support equipment a complex system is likely to require. Support equipment includes tools, TMDE, condition-monitoring equipment, metrology, and calibration equipment, computer- and software-support tools and equipment, maintenance fixtures and stands, fueling and arming equipment, battle damage assessment and repair (BDAR) tools and equipment, and special handling equipment required to support maintenance at every level. A system designed to use common or existing support equipment, while limiting unique or special equipment, helps reduce the logistics burden on the warfighter.

TECHNICAL DATA. Technical data include installation, operation, and maintenance instructions; inspection and calibration procedures; overhaul instructions; computer or software debugging and recovery instructions; "safe mode," degraded operations, and limited-functionality use procedures; exception-handling and fault-tolerance instructions; facilities data; modification instructions; engineering data (specifications, drawings, software development data, materials, and parts lists); supplier data, and logistics provisioning and procurement data that are necessary for system development, production, operation, maintenance, and disposal functions. Procuring the *necessary* engineering data requires significant analysis of the potential system life cycle, data rights, potential diminishing manufacturing sources and material shortages (DMSMS) challenges, reprocurement possibilities, desires for repair parts and spares break-out (development of alternative sources), and potential data repositories. Data procurement is expensive and should be limited to elements with clear immediate or potential value to the system.

TRAINING AND TRAINING SUPPORT. Training and training support includes all personnel, equipment, facilities, ranges, data or documentation, programs of instruction, and associated resources for the training of the operational, maintenance, and support personnel at every level and through every phase of system development. Training resources, including simulators, mock-ups, embedded training systems, and special devices necessary to support new equipment training (NET), qualification, familiarization, safety, and formal and informal training are also included.

COMPUTER RESOURCES SUPPORT. Computer-resources-support planning, dominated by the software support component, will continue to grow as developed systems continue to become more software intensive and software dependent and as net-centric systems are developed and deployed. As already discussed, the immaturity of the software engineering environment and the high cost of software maintainers, combined with the tendency towards vagueness in high-level, performance-based specifications often result in software designs that are difficult and costly to sustain. Computer resources support includes computers; networks; software [operating, security, database, prognostic/diagnostic, support, middle-ware, interfacing/interoperability, training, modeling and simulation (M&S), and other associated software]; and computer-aided software-engineering (CASE) tools, subscriptions, and media necessary to support scheduled and unscheduled activities at every level. Driving the software architectural design towards efficient and cost-effective sustainability is essential for software-intensive TOC and system performance.

FACILITIES. This category includes physical plant, portable buildings, erectable structures/shelters, mobile vans/trailers, revetments, hangars, docks/boat yards, calibration laboratories, software integration laboratories (SILs), and special repair shops (depot, overhaul, clean rooms, etc.) necessary for system sustainability. Capital equipment and utilities are generally included in facilities support planning.

PACKAGING, HANDLING, STORAGE, AND TRANSPORTATION (PHS&T). PHS&T includes all materials, equipment, special provisions, disposable and reusable containers, data repositories, and supplies necessary to support the packaging, preservation, storage, handling, and transportation of all mission-related elements of the system. Transportation modes considered include air, surface (highway, rail, waterway, etc.), and pipeline.

DESIGN INTERFACE. As with any system performance attribute or requirement, the realized level of supportability performance is linked to the system design. Simply stated, the performance is determined by the design, so that the design must be influenced to achieve the performance needed. The fact that the race car can be refueled and have four tires replaced in a few seconds is no accident—that performance attribute is meticulously designed into the car's architecture. Tradeoffs and compromises are considered against a high standard—*will the design help win the race?*

The system supportability performance must be a design interface and held to a similar high standard—*will the design maximize the system's combat capability for the warfighter*? System supportability design is critical to effectiveness, suitability, and TOC. Decisions and tradeoffs made during early concept exploration and design analysis significantly influence sustainability design, downstream costs, and combat power. The following section describes the sustainability design interface in more detail.

6.2 *DESIGN FOR LOGISTICS PERFORMANCE*

6.2.1 *SUPPORTABILITY ANALYSIS*

The purpose of supportability analysis is to optimize the system providing combat power to the warfighter and to do so while minimizing the TOC. Logistics supportability has always been a prominent element of suitability assessments, but it is often overlooked as a system performance element in effectiveness assessments. Conducting supportability analyses as a performance element that impacts both system effectiveness and suitability will result in the development of a better performing system.

Supportability analysis is an essential part of the SEP and begins early in a system's concept development. It utilizes the initial capabilities document (ICD), known constraints, market analyses and surveys, CAIV and tradeoff analyses, and the proposed system's doctrine, organization, training, material, leadership, personnel, and facilities (DOTMLPF) as SEP "inputs." In the initial SEP iterations, the "output" would include a complete set of stated, derived, and implied system supportability requirements. Subsequent SEP iterations should produce the system supportability architectural design, reviewed at the preliminary design review (PDR); supportability production design, reviewed at the critical design review (CDR); and eventually, the actual system supportability elements for testing, acceptance, and production. The government user- and program-management communities together must develop the supportability requirements, effectively communicate those requirements through the RFP to potential contractors, and drive the design for supportability through technical and design reviews, audits, testing, and assessments. Some identified supportability analyses tools and methodologies are identified in Table 6.1.

TABLE 6.1 SUPPORTABILITY ANALYSIS TOOLS [3]

Analysis tools	Description of application
Life-cycle cost analysis (LCCA)	Analysis of the system life-cycle cost (research, development, management, testing and evaluation, production, sustainment, and disposal costs); considers high-cost contributors, cause-and-effect relationships, potential areas of risk, and identification of areas for improvement; serves as input for CAIV and tradeoff analyses.
Maintainability, upgradeability, interoperability/interfaces, reliability, and safety & security (MUIRS) analysis	Software-oriented supportability analyses to more fully develop stated, derived, and implied requirements, compensating for the state of relatively immature software engineering in a performance-based environment; the purpose of these analyses is to elicit proposals with a higher degree of cost and schedule realism and to drive the software architectural design towards improved supportability performance and lower TOC.
Level of repair analysis (LORA)	Evaluation of maintainability concepts, maintenance policies, logistics footprint, and system supportability attributes to determine the number of levels necessary; analyzes reparability options, tasks, skills, and tools and equipment for tasks assigned to each level; considers economic, technical, training, and environmental factors.
Maintenance task analysis (MTA)	Evaluation of maintenance functions and tasks in terms of task elapsed time, sequences, personnel needed to perform, skill levels, and supporting resources necessary (e.g., spares/repair parts and associated supplies, hardware and software tools and test equipment, facilities, transportation and handling requirements, technical data, training, and computer resources); identification of high resource-consumption areas.

(*Continued*)

TABLE 6.1 SUPPORTABILITY ANALYSIS TOOLS [3] (CONTINUED)

Analysis tools	Description of application
Reliability-centered maintenance (RCM)	Evaluation of the system in terms of the life cycle, to determine the best overall program for preventive (scheduled) maintenance; the focus is on driving the system design towards affordable system reliability and on establishing a cost-effective preventive maintenance program based on reliability information derived from the FMECA.
Fault-tree analysis (FTA)	A deductive approach involving the graphical enumeration and analysis of differing fault and failure modes and the probability of occurrence; fault trees can be developed for identified failure modes or undesired events, such as system software degradation/slowdown. The focus is on an identified top-level event, and the first tier causes associated with it. Each of these is further investigated for their causes and so on. The FTA focus is narrower than a FMECA analysis and does not require as much input data.
Failure modes, effects, and criticality analysis (FMECA)	Analysis and identification of potential product or process failures, the expected failure modes, causes and mechanisms, failure effects on the system/mission, anticipated frequency, criticality, and possible mitigation (e.g., engineering redesign, preventive measures, revised or added maintenance action, etc.).
Evaluation of design alternatives	An overall evaluation of alternative design configurations for optimizing system supportability within cost and schedule parameters; balances system supportability with other system performance parameters to maximize the system's combat capability to the warfighter.

6.2.2 MUIRS ANALYSIS TECHNIQUE FOR SOFTWARE-INTENSIVE SYSTEMS [4]

As presented earlier, those performing system supportability planning must recognize that the software engineering environment is significantly immature when compared to most hardware-centric engineering environments, and that lack of maturity must be compensated for through comprehensive requirements and architectural design development. To produce the system capability desired, the software developer must have significant knowledge of the requirements for MUIRS performance attributes. In addition, the developer must also know the operational context or scenarios those attributes operate within to design an effective software architecture.

MAINTAINABILITY. The software engineering environment has few, if any, maintainability standards that are widely accepted or normally practiced in software design and development. In contrast, the mature automotive engineering environment has many such standards and practices—such as provisions for replacement of oil, filters, tires, batteries, and other components that are likely to require maintenance over the life of the system. When considering what requirements need to be communicated to a competent automotive manufacturer, an end user need not specify any maintainability attribute that is an industry-wide standard or common practice, as those requirements are considered to be implied. The software engineer has no industry-wide standards or common practices for software maintainability design, so that detailed requirements and architectural design management are necessary. In short, the software developer will respond to requirements provided and design guidance almost exclusively—a customer will get the maintainability design he/she asked for and very little else. Stated, derived, and implied software maintainability requirements must be developed and provided to the software engineers, and design reviews must be conducted to ensure that the system maintainability performance is achieved.

Upgradability. Like maintainability, upgradeability is a design-oriented attribute. The immature software engineering environment creates the same type of problems in the design for upgradeability as it does for maintainability—the lack of industry-wide standards means that requirements for upgradeability must be fully detailed and effectively communicated to the software developer. All known, planned, and reasonably foreseeable system upgrades must be identified in as much detail as possible so that the software architecture can be designed to accommodate upgrades with the minimum amount of reengineering work. The relative cost for designing the software to accommodate upgrades is extremely low when compared to the effort required for reengineering unplanned upgrades. Thus, including upgradeability design, even if only one of many planned upgrades is accomplished, is much more cost effective. Identifying as many known, planned, and foreseeable upgrades as possible is a key component in designing a system with an open-architecture (OA) focus.

Interoperability/Interfaces. The system's design for interoperability and interfaces is another key component in its OA attributes, especially for SoS and net-centric systems. The immature software engineering environment creates the same nonsupportive development environment and drives the same level-of-requirements definition as the maintainability and upgradeability attributes just described. There are few industry-wide standards to enable interoperability, so that design decisions are key in obtaining the desired level of interoperability and interface performance. Similarly, all known, planned, and foreseeable interoperability and interfacing requirements must be specified and sufficiently detailed to enable the software engineer to design an accommodating, open oriented architecture.

Reliability. Modern systems can well have more than 80% of their system functionality performed by software; thus, software reliability is the dominant element in these systems' reliability performance. As with any system, reliability must be designed into the development process and software product. The requirement for system reliability will mandate that system component reliability, including software components, be very high so that system reliability thresholds are met. For example, a system with three components, all with 95% reliability, arranged in series provides a system reliability of only 85.7% ($0.95 \times 0.95 \times 0.95 = 0.857$). Often, redundant components arranged in parallel must be designed to meet system reliability. The component reliability in a redundant design is one minus the probability of all redundant systems failing. Using the same example with one of the 95% reliable components duplicated and arranged in parallel (redundant capability design), the system achieves a 90% reliability [$0.95 \times (1 - 0.05 \times 0.05) \times 0.95 = 0.900$]. Microsoft's "Safe Mode" operating system illustrates the concept of a redundant software component: a problem with the primary operating system triggers the Safe Mode redundant operating system. Although effective at increasing reliability, redundant systems can also increase the maintenance burden as failed components must still be serviced, maintained, or replaced to ensure system reliability is not degraded. Effective communication of system reliability requirements is another essential element in driving the software developer to a design for effective and suitable warfighter capabilities.

Safety/Security. Safety and security software design elements might well be matters of life and death to the warfighters using the system, and their importance cannot be overstated. Software safety is linked with reliability, as more reliable systems are inherently safer. Software-oriented system safety components tend to be more abstract than hardware-oriented safety elements; therefore, effective communication of safety-related performance is necessary. Redundant, Safe Mode operations for essential operations such as avionics, weapons employment, navigation, etc., must be well understood by the developer to ensure that the design accommodates those safety-related functions.

Likewise, security of the system is critical to the warfighter, especially in net-centric warfighting systems. Like any other warfighting system, the enemy will seek vulnerabilities and actively attack any security weaknesses with very sophisticated methods. The need for system security influences the system's entire design. And, as it is nearly impossible to add effective security to an existing software design, system security requirements must be well communicated to the software developer before software design work begins.

Many aspects of system safety and security, including software components, will be illuminated through the FMECA technique described next.

6.2.3 FAILURE MODES AND EFFECTS CRITICALITY ANALYSIS

Failure modes and effects criticality analysis is an essential tool in shaping both the hardware and software architectural design. The design engineers must be able to evaluate a system with a view towards possible failures, the anticipated modes of failure, expected frequency, causes, consequences, and impacts of such failure, and potential preventive maintenance actions or reengineering opportunities that would likely mediate identified failures. The objective is to identify failure-prone components, processes or elements, the criticality and impact of the failures on the system/mission, potential safety or software configuration item issues, and areas of high risk.

Net-centric and other software-intensive systems are dependent on the system software for ever-increasing portions of system functionality. Consequently, software-oriented FMECA is critical in driving a system design that is reliable, sustainable, operable under extreme and degraded modes, and recoverable when failures occur. Orienting FMECA on the software architecture is critical in achieving system reliability, safety, and operations in degraded modes. Too often, software systems fail because of relatively unimportant system inputs or minor discrepancies (because the designers were unable to separate essential routines from enhancing routines), and so a failure anywhere tends to be critical. For example, when a flight of 12 F-22 Raptors crossed the International Dateline for the first time, all software functions failed except for those required for flight surface control. The change in date and time is fairly inconsequential to the combat capability of the F-22, but that unrecognized input caused a critical system failure. In other words, the software developers were unable to separate nonessential inputs to ensure that unexpected input values could be immediately rejected without impact, and certainly without critical impact, to the Raptor's combat capability. A software patch—reengineered code—was provided to fix the problem.

FMECA consists of a nine-step process [3]:

1) Identify the different mission and system failure modes.

2) Determine the cause(s) of failure.

3) Determine the subsequent effects of the failure on other systems/processes.

4) Identify failure detection and verification methods.

5) Determine the frequency of occurrence.

6) Assess the severity of the failure.

7) Determine the probability that a failure will occur and be detected.

8) Analyze the failure mission or system criticality based on the preceding steps: severity of the failure and the probability of occurrence and detection.

9) Identify critical areas and recommend mitigation alternatives, including revised processes, preventive maintenance actions or revisions, or reengineering opportunities.

6.2.4 Reliability Centered Maintenance

Reliability centered maintenance (RCM) determines actions that must be completed to ensure that any system continues to fulfill its mission as defined by the user/warfighter. Similar to FMECA, the purpose of RCM is to design system maintenance tasks and actions that are focused on system reliability and to identify the need for reengineering for critical failure modes that cannot be addressed through maintenance. There are seven basic questions that focus the RCM analyses [5]:

1) What are the functions and associated performance standards of the system in performing its mission?
2) In what ways does it fail to fulfill its functions?
3) What are the causes of the functional failure?
4) What is the effect of each failure?
5) How critical is each failure?
6) What can be done to prevent or predict each failure?
7) What should be done if a suitable preventive or mitigating task cannot be implemented?

Although this process sounds very logical, a surprising number of maintenance tasks are established without a RCM-based analysis. For example, a daily preventive maintenance task for virtually every one of the approximately 500,000 vehicles in the U.S. Department of Defense (DoD) was, in the not too distant past, "check batteries daily." If an RCM analysis were conducted, it would suggest that there was some battery-related, critical failure that would occur, with high probability, within one day's normal operation AND that the battery check would likely expose the impending failure so that it could be prevented, thus improving the reliability of the systems. This might sound trivial until one analyzes the manpower expended for this effort: Consider that the battery check would take, on average, one minute for each system and that each system was operational about 250 days per year, the effort devoted to this effort would be $(500{,}000 \times 1 \times 250)/60 = 2{,}083{,}333$ man-hours. In addition, the battery checks themselves might have caused failures as the cables were flexed, postconnections were twisted, and contamination was introduced when filling cells on maintainable batteries.

From the preceding example, it is clear that RCM analyses require an understanding of how components are likely to fail over the life cycle of the system so that periodic maintenance tasks are highly leveraged. Modern systems are much more complex than in the past, and components have widely differing wear-out patterns. The graphs in Fig. 6.5 depict the varying wear-out patterns identified in modern systems and the percentage of modern systems that exhibit failures associated with that pattern.

These six graphs represent the differing failure and wear-out patterns prevalent in modern systems, with the percentage of systems exhibiting the corresponding failure pattern shown to the right of each. Graph A depicts the well publicized "bathtub curve" failure pattern, in which there are a number of initial failures (components experiencing infant-mortality failure) followed by a stable period of relatively low failure and culminating in an increasing failure rate as components fail from age and wear. The bathtub-curve failure pattern was widely believed to be the most common, but, as indicated in Fig. 6.5, only 4% of modern systems exhibit this type of wear-out pattern. Graph B, representing only 2% of systems, depicts a stable failure rate, culminating in a similar increase in failure rate from age and wear. Graph C, representing 5% of systems, depicts a slowly increasing failure rate with no identifiable rapid wear-out stage. Graph D, representing 7% of systems, indicates a very low initial failure rate, rising to a constant, stable rate over the remaining system life. Graph E, representing 14% of systems, depicts a stable, random failure rate that remains essentially unchanged throughout the system life cycle. Graph F, representing 68% of systems, depicts a high initial failure rate rapidly declining to a stable, or continuously decreasing, failure rate over the remaining system life. Although software does not wear out in the traditional sense, it is a significant driver for system failures and maintenance actions. System software is likely to be most aligned with graph F, as test programs typically reveal between 50 and 75% of the software errors, and a significant number of maintenance actions are usually required immediately after deployment, and then decline rapidly.

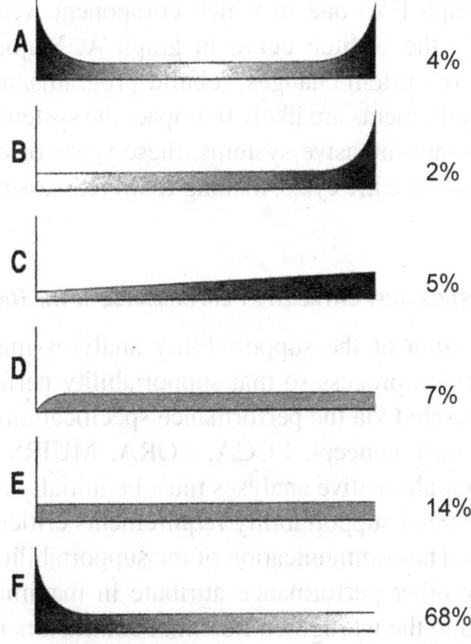

Fig. 6.5 Patterns of failure and wear-out [5].

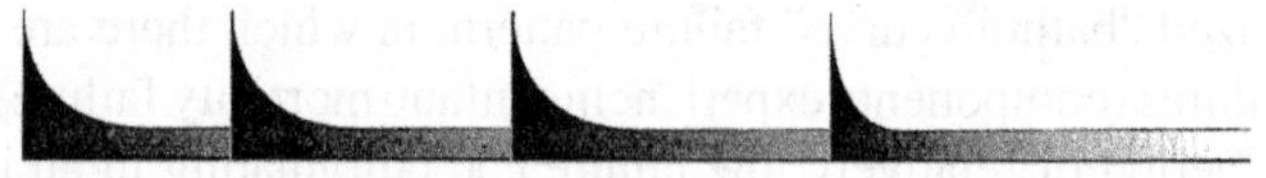

Fig. 6.6 Software-intensive system failure pattern [6].

Although graph F is typical for software component initial deployment, the failure rates and maintenance requirements tend to spike up as major system changes impact the software, such as software upgrades, rehosting software on new computers or hardware, new interface or interoperability requirements, safety and security updates, etc. For software-intensive systems like net-centric warfighting platforms, the software failure and maintenance requirements can dominate the system's failure pattern. This spiking effect tends to produce a saw-blade-type pattern, with both predictable and unpredictable, sudden increases in failures and required maintenance actions. This pattern is depicted in Fig. 6.6.

To achieve true RCM, the system's supportability concept must be aligned with the failure pattern expected, matching preventive and corrective maintenance resources to address the highest probable failure periods. It is important to understand that these patterns might not be static over a system's life cycle; periodic reevaluation of the system must be accomplished to maintain the RCM approach. This is especially true when there are significant changes to the system or the intended life cycle. DoD routinely operates systems far beyond their intended life cycles, which could change a system's pattern from that depicted in graph F to one in which component wear-out affects the failure rate, that is, the bathtub curve in graph A. Major mission profile changes, upgrades or system changes, rebuild programs, added or changed interoperability requirements are likely to impact the system's failure pattern. In the case of software-intensive systems, these types of changes typically occur often throughout the life cycle, making RCM reassessment a continuous, ongoing process.

6.2.5 *Driving Logistics Supportability Performance in the RFP*

A significant portion of the supportability analyses must be completed early in the acquisition process so that supportability performance requirements are communicated via the performance specifications in the RFP. The system's supportability concept, LCCA, LORA, MUIRS analyses, RCM, FMECA, and design-alternative analyses must be initially conducted to capture derived and implied supportability requirements critical to the proposal and design process. The communication of the supportability requirements is as essential as any other performance attribute in maximizing the combat capability provided to the warfighter. Potential contractors must have a clear picture of the supportability requirements to accurately estimate the level of effort and provide a more realistic proposal. Discovering derived and implied supportability requirements after the RFP is issued will likely result in a misinterpreted level of effort and a significantly understated proposal in terms of cost and schedule. If they are discovered after the contract is let, much initial work will likely be scrapped and reworked, negatively impacting the program cost and schedule. The worst-case scenario is that derived and implied supportability requirements are discovered in IOT&E, increasing the logistics burden (in the best case) or rendering the system operationally ineffective or unsuitable (in the worst).

6.2.6 LOGISTICS SOURCE-SELECTION CONSIDERATIONS

The supportability analyses conducted will typically identify key system supportability attributes that can be designated as KPPs or are sufficiently important to use as source-selection discriminators. This systems engineering approach reflects the goal of *maximizing system performance* and *optimizing the combat capability* for the warfighter. As with any other source-selection criteria, the relative importance of system supportability to the potential contractors' resulting proposals will likely emphasize the potential vendors' approach to meeting the supportability criteria specified. In addition, the system's supportability attributes profoundly impact the TOC, so that TOC performance is significantly influenced by the source-selection supportability criteria selected.

6.2.7 LOGISTICS TEST AND EVALUATION

Testing and other verification methods are planned to verify that the system meets the performance specified, but it is often difficult to design verification methodologies for much of the supportability characteristics. This is especially true for software-intensive systems, in which much of the design effort is focused on future needs such as upgradeability, interoperability, and interfacing. Testing is expensive and time consuming, and so the number of tests and duration of testing are under constant pressure for reduction. It is unlikely—and undesirable—that any system would experience the failures and conditions necessary to demonstrate all of its critical supportability attributes during scheduled testing, and it would surely fail to meet test performance requirements if it did. Specific and highly leveraged supportability tests, demonstrations, modeling, and simulation must be planned throughout the contractor and government test and evaluation (T&E) programs. Developing and tracing supportability requirements to operationally representative scenarios and test cases (which include demonstrations, modeling, and simulation) is an effective way to build comprehensive supportability test programs to augment supportability data that might or might not be garnered from other operational testing. It is important to ensure that the supportability performance desired has corresponding verification methodologies in the overall contractor test plan and test and evaluation master plan (TEMP).

6.3 ADVANCED LOGISTICS CONCEPTS

6.3.1 SUPPLY-CHAIN MANAGEMENT

Supply-chain management [7] refers to the network for sourcing raw materials, to manufacturing products or creating services, to storage and distribution of goods, and to ultimately delivering the products or services to the customers and consumers. This network includes both internal and external elements of concern, focusing on efficiencies that satisfy customer needs and expectations, including cost efficiencies. Modern information technologies, product-tracking methods, transportation systems, and forecasting techniques have enabled effective analyses of complex supply chains and have provided the potential for advanced techniques to improve supply performance, to improve efficiency, and to lower cost. DoD is interested in advanced supply-chain management as both a consumer of products and services offered and a provider of products and services to the ultimate customers: the citizens and government of the United States.

The purpose of advanced supply-chain management is to improve a total system of supply so that performance improvements and efficiency savings are accrued and shared across the network. Entities that successfully implement supply-chain management strategies and techniques are better able to compete for DoD business through responsive and lower-cost supply-chain operations. DoD must also be a responsive supplier to the warfighter and must leverage scarce funding resources for maximum efficiency; effective supply-chain management concepts and techniques offer the potential for both.

There are four levels of supply-chain optimization, as depicted in Table 6.2.

The four levels illustrated depict a progression from basic through advanced supply-chain management, with the first two levels grouped as "internal" and the last two levels grouped as "external." Nearly all of the agencies implementing supply-chain optimization transition from one level to the next, and there is no practical way to jump to higher levels. Arrayed along the vertical axis are key elements in supply-chain management that assist in describing players, benefits, and attributes at each level.

6.3.2 CONTRACTOR LOGISTICS SUPPORT

In nearly every war or conflict in which the United States has been involved, there has been some form of contractor support used in the effort, including on the battlefield and at sea. Although not new, the role of contractor logistics support (CLS) has significantly expanded into more and more supportability

Table 6.2 Levels of Supply-Chain Optimization [7]

	Internal		External	
	Sourcing and logistics I	Internal excellence II	Network construction III	Industry leadership IV
Driver	VP sourcing (under pressure)	CIO/supply chain leader	Business unit leaders	Management team
Benefits	Leveraged savings	Prioritized improvements across network	Best partner performance	Network advantage, profitable revenue
Focus	Inventory, logistics, freight, order fulfillment	Process redesign, system improvement	Forecasting, planning, customer services, interenterprise	Consumer, network
Tools	Teaming, functional excellence	Benchmarks, best practice, activity based costing	Metrics, database mining, electronic commerce	Intranet, internet, virtual information systems
Action area	Midlevel organization	Expanded levels	Total organization	Full enterprise
Guidance	Cost data, success funding	Process mapping	Advanced cost models, differentiating processes	Demand-supply linkage
Model	None	Supply-chain—interenterprise	Interenterprise	Global market
Alliances	Supplier consolidation	Best partner	Formal alliances	Joint ventures
Training	Team	Leadership	Partnering	Network processing

functions, including supply and maintenance support embedded within tactical units and warfighting ships. Software-intensive systems, high levels of technology, improved reliability, competition with industry for talent and skilled personnel, outsourcing of logistics "tail" (depots, base support, messing operations, etc.), and downsizing of the armed forces have all contributed to the increase in CLS. The training and education of uniformed military in supporting software-intensive and other complex systems is, in many cases, cost prohibitive or infeasible; therefore, CLS approaches are likely to continue to be a significant and important element in systems' supportability concepts and analyses.

When developing a system's supportability concept, analyses should include functions designated for CLS, level of support, whether the contract is for initial support only or life-cycle support, support scheme during combat and open conflict, and associated risk-management functions. For example, it is unlikely that software would be fully supported by organic, uniformed personnel; thus, a net-centric warfare system will likely require a significant CLS structure. This structure must be included in the analyses and planned in detail to ensure adequate support of that critical asset.

6.3.3 PERFORMANCE-BASED LOGISTICS (PBL)

PBL, as defined by DoD Directive 5000.1, "is the purchase of support an integrated, affordable, performance package designed to optimize system readiness and meet performance goals for a weapon system through long-term support arrangement with clear lines of authority and responsibility" [8]. Although this definition is clearly focused on contracted supportability performance, the concept of utilizing PBL capabilities to optimize system performance within cost and other constraints is key in maximizing combat capability for the warfighter.

A PBL strategy focuses weapon system support on identified warfighter-required performance outcomes, rather than on discrete transactional logistics functions. It should balance three major objectives throughout the life cycle of the weapon system: deliver sustained materiel readiness, minimize the requirement for logistics support through the incorporation of reliability-enhancing technology insertion and refreshment, and improve the cost effectiveness of logistics products and services.

Conceptually, these PBL objectives should apply to the entire system-supportability concept, not just that of a contracted solution. In addition, a PBL approach is suitable for new development, existing, and legacy systems.

As with any performance-based attribute analysis, it is critical to understand the supportability requirements in great detail (using a systems engineering approach) and to perform the necessary supportability analyses to determine key performance outcomes. The challenge in communicating and achieving supportability performance outcomes for software systems or components is even greater. Contracting for PBL entails a 12-step process [8]:

1) Integrate requirements and support.
2) Form a PBL team.
3) Baseline the system.
4) Develop performance outcomes.
5) Select the product support integrator.
6) Develop a workload allocation strategy.
7) Develop a supply-chain management strategy.
8) Perform a PBL support strategy analysis.
9) Establish performance-based agreements.
10) Award contracts.
11) Employ financial enablers.
12) Implement and assess.

All of the concepts presented in this chapter are applicable in performing PBL analyses on any system. Failure to consider supportability as integral to the system being developed will likely result in a suboptimized system that robs the warfighter of combat capability.

6.4 Conclusion

Optimizing total systems performance requires a systems-engineering approach to developing the required combat capability specified in the JCIDS documents. This systems approach means that all performance attributes, including supportability performance, are considered equally when performing tradeoff analyses to maximize *warfighter effectiveness and suitability* within known constraints. One of those constraints is nearly always TOC, and supportability performance is a major TOC driver in the system.

The system architectural design is one of the keys to attaining desired supportability performance and effective supportability analyses conducted early in the concept development are critical in shaping that design. Resulting supportability performance requirements must be clearly articulated to potential system developers in the RFP and source selection processes, and the design process must be continuously assessed for supportability performance. Supportability testing and other verification methods must be planned into the TEMP and contractor test plans.

System supportability and TOC performance are significantly locked in early in the system's concept and design phases. Significantly changing this performance after the system is designed is nearly impossible without a major costly and time-consuming reengineering effort.

References

[1] General Accounting Office, "Best Practices: Setting Requirements Differently Could Reduce Weapon Systems' Total Ownership Costs," GAO-03-57, U.S. Government Printing Office, Washington, D.C., Feb. 2003.

[2] Chairman of the Joint Chiefs of Staff, "Joint Capabilities Integration and Development System," Chairman of the Joint Chiefs of Staff Instruction 3170.01F (CJCSI 3170.01F), U.S. Government Printing Office, Washington, DC, May 2007, p. B.3.

[3] Blanchard, B., *Logistics Engineering and Management*, 5th ed., Prentice-Hall, Upper Saddle River, NJ, 1998, pp. 185–187.

[4] Naegle, B., *Developing Software Requirements Supporting Open Architecture Performance Goals in Critical DoD System-of-Systems*, Naval Postgraduate School, Monterey, CA, 2006, pp. 17–25.

[5] Moubray, J., *Reliability-Centered Maintenance*, 2nd ed., Industrial Press, Woodbine, 1997, pp. 7–16.

[6] U.S. Air Force Software Technology Support Center, *Guidelines for Successful Acquisition and Management of Software-Intensive Systems (GSAM)*, ver. 3, Hill Air Force Base, UT, 2000, p. 12–8.

[7] Poirier, C., *Advanced Supply Chain Management*, Berrett-Koehler, San Francisco, 1999.

[8] Defense Acquisition University, *Defense Acquisition Guidebook*, 2004, http://akss.dau.mil/dag/DoD500.asp?view=document [retrieved 1 March, 2007].

STUDY QUESTIONS

6.1 How is logistics supportability linked with system performance?

6.2 How important is the acquisition process in determining logistics supportability performance and TOC?

6.3 In which phase of a system's life cycle is the TOC performance determined?

6.4 How does the degree of software intensity impact a system's supportability aspects?

6.5 What are the 10 logistics supportability elements, and when in the process should they be considered?

6.6 What is the purpose of logistics supportability analysis?

6.7 Why is the MUIRS analysis methodology a necessary and effective tool for software-intensive system LSA?

6.8 How does supply-chain management impact system supportability and TOC?

6.9 What is reliability-centered maintenance, and how is it applied?

6.10 Explain the concept and applicability of performance-based logistics.

Chapter 7

Test and Evaluation Management

Brad Naegle* and Keith F. Snider†
Naval Postgraduate School, Monterey, California

Learning Objectives

- Understand the purposes of T&E
- Distinguish between DT&E and OT&E
- Understand the difference between operational effectiveness and operational suitability
- Understand major T&E legislation
- Describe the purposes of LFT&E
- Describe how the DoD organizes for T&E
- Understand the contributions of program documents in T&E planning
- Describe the role and content of the TEMP
- Understand how statistical methods are used in T&E and limitations of those methods
- Describe the role of modeling and simulation in T&E

7.1 Introduction

Test and evaluation (T&E) in acquisition program management refers to activities conducted throughout the project life cycle to assess an item's performance, to reduce risk, and to provide information for decision makers. Strictly speaking, the word *test* denotes an event conducted to gather data, whereas *evaluation* means the analysis of data that results in useful information. Other terms that often arise in the realm of T&E include *verification*, which refers to proof that an item complies with its specifications ("Was it built right?"), and *validation*, which is proof of customer satisfaction ("Was the right item built?").

*Senior Lecturer, Graduate School of Business and Public Policy.
†Associate Professor, Graduate School of Business and Public Policy.

7.2 PURPOSES OF T&E

T&E is conducted for several reasons:

1) *It is an integral step in the systems-engineering process* (SEP), as it verifies performance, detects deficiencies, and validates requirements. T&E, along with other activities such as inspections, demonstrations, modeling and simulation, accomplishes the verification and validation functions of the SEP. From a system engineer's perspective, T&E is embedded in the SEP. As the SEP proceeds iteratively through the project life cycle, T&E serves to assess the project's technical maturity. It measures the effectiveness of progressively evolved products and processes, as well as the extent to which specifications for those products and processes are met. Results from T&E activities are key inputs at project technical and design reviews. In essence, then, effective SE requires effective T&E (Fig. 7.1).

2) *It is an essential element in acquisition decision making.* T&E provides information to support tradeoffs, refine requirements, and manage risk. Although the SEP makes up the technical aspect of the project life cycle, decision makers must consider other aspects, such as the project's business case and the available budget [1]. Throughout the system's life cycle, but particularly at each milestone review, T&E activities provide decision authorities with critical information concerning the project's technical progress, which obviously affects its cost and schedule.

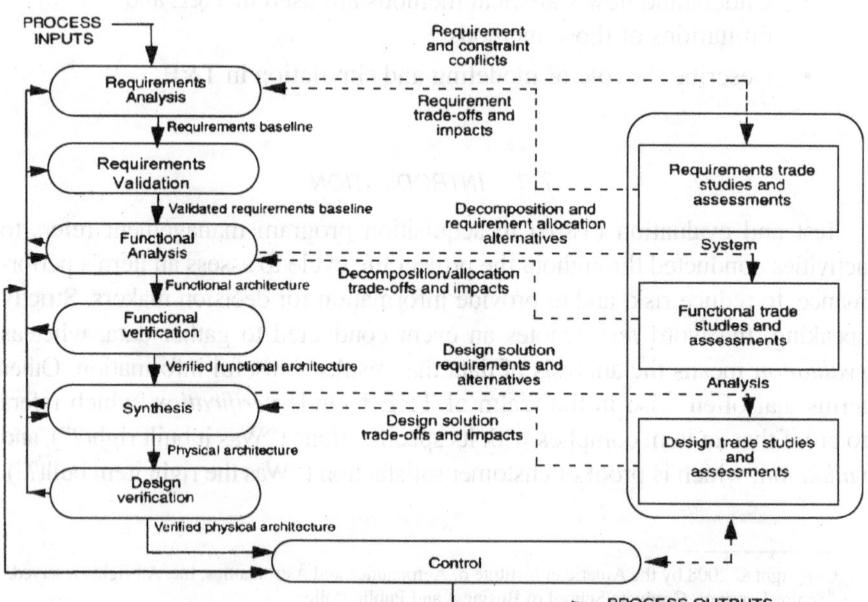

Fig. 7.1 T&E as verification and validation in the systems-engineering process [2].

3) *It is required by law*. The most significant T&E-related legislation concerns operational T&E (OT&E) and live-fire T&E (LFT&E), each described further in the following sections. In many cases, T&E legislation reflects both Congress' dissatisfaction with the adequacy of the DoD's and contractors' past testing and its corresponding desire to exercise more control to ensure adequate testing in the future.

7.3 TYPES OF T&E

Developmental T&E (DT&E) and OT&E constitute two broad categories of T&E in defense acquisition. In a very general sense, DT&E addresses the verification steps of the SEP, whereas OT&E concerns validation. Within these two categories, several specific types of T&E can be identified. Further, some specialized types of T&E are necessary in various programs.

7.3.1 DT&E

DT&E encompasses engineering types of tests that serve, among other purposes, to substantiate technical performance, verify design risks, certify system and component safety, and assess readiness for OT&E. It generally is highly controlled, requires significant measurement through instrumentation, and is performed by technical rather than operational personnel. Especially at the component and subsystem level, DT&E often entails repetitive test events on single test articles to attain some statistical significance of results. DT&E is performed throughout the system life cycle at the component, subsystem, and system level (Fig. 7.2).

Areas of DT&E emphasis include performance; logistics supportability; reliability, availability, and maintainability; manpower and personnel integration; safety verification; environmental protection; survivability; and interoperability.

a) b)

Fig. 7.2 DT&E: a) at the subsystem level—radar in an anechoic chamber; and b) at the system level—tractor and tank on a tilt table.

Among the many forms that DT&E takes, major types include the following:

- *Qualification testing* can be conducted for two purposes: 1) to assess a product's compliance with requirements in order to approve the vendor as a source of supply; or 2) to assess an item or system design within particular environmental operating conditions to determine whether it can function within certain safety factors (as in qualifying an aircraft engine to operate under specified flight conditions).
- *Production qualification testing* assesses manufacturing and production processes, facilities, and equipment. It samples initial production items and is typically conducted before a program is approved to enter full-rate production.
- *Production acceptance test and evaluation* assesses manufactured items to demonstrate that they satisfy requirements.

Because they perform almost all system development and production tasks, contractors perform most DT&E. However, the government maintains significant DT&E capabilities at major test facilities such as Aberdeen Proving Ground, Edwards Air Force Base, and Naval Surface Warfare Center. DT&E activities at these government facilities are usually performed at the major subsystem and system levels.

7.3.2 OT&E

OT&E is conducted throughout the system life cycle to determine if operational requirements have been met and to assess operational effectiveness ("Does it work?") and suitability ("Does it fit?").

For example, a rifle's range, muzzle velocity, and accuracy are effectiveness parameters, whereas its weight and ease of maintenance are suitability parameters.

To assess whether such operational considerations are met, OT&E activities are conducted in as realistic conditions as possible, including production-representative test articles, typical operators, representative targets and threats, realistic environments and scenarios, force-on-force engagements, and restricted contractor participation.

Legislation requires both realistic OT&E for a major program and reporting on that testing before the program can enter full-rate production.

There are several types of OT&E:

- *Operational assessment (OA)*: OT&E conducted during system development phases, usually on prototypes or other preproduction articles
- *Initial OT&E (IOT&E)*: dedicated OT&E that satisfies the statutory requirement for realistic OT&E before the project can enter full-rate production (This term is misleading because OAs are conducted prior to IOT&E.)
- *Follow-on OT&E (FOT&E)*: OT&E conducted after fielding

7.3.3 Combined Testing

DT&E and OT&E can be combined when stakeholders agree that it makes good business sense to do so. For major systems, however, some dedicated portion of IOT&E will be necessary in order to satisfy the statutory requirement for realistic testing prior to full rate production.

7.3.4 Specialized T&E

LFT&E assesses two system characteristics: *vulnerability*, damage caused to a system and its crew when engaged by threat weapons, and *lethality*, damage inflicted by a system on typical threat targets. (*Vulnerability* and *susceptibility* are two variables that define a system's *survivability*. Susceptibility refers to the extent to which a system can be engaged by threat weapons, whereas vulnerability refers to damage to the system once it has been engaged.)

A system's LFT&E program encompasses vulnerability and lethality assessments on components, subsystems, and systems throughout its life cycle. Though such T&E should occur as part of the SEP to ensure vulnerability and lethality requirements are met, Congress has enacted legislation requiring the DoD to conduct LFT&E on certain systems. [The statutory requirement applies to "covered systems" (defined as systems that provide some measure of crew protection), to major munitions programs, and to programs that will procure ammunition in large quantities.] One of the most important statutory provisions is the requirement for vulnerability testing on operationally configured (i.e., with full fuel and ammunition loads) production-representative systems and for a report on that testing prior to the full-rate production decision. This requirement might be waived under certain conditions.

LFT&E has the character of DT&E in that it is carefully controlled and highly instrumented (Fig. 7.3). It is also usually augmented heavily by modeling and simulation.

Other specialized types of T&E include multiservice T&E; joint T&E; nuclear, biological, and chemical weapons T&E; nuclear hardness and survivability T&E; and space system T&E.

7.4 Organization for T&E

Authorities and responsibilities for T&E are separated along the boundary between DT&E and OT&E.

DT&E, as the activity that ensures that systems are built correctly (i.e., according to requirements and specifications), falls completely under the purview of the project/program manager (PM). Almost all DT&E is

Fig. 7.3 LFT&E for vulnerability.

accomplished either by contractors or by government test facilities, both under the direction and oversight of the PM. DoD and service oversight of DT&E is accomplished by staff representatives in the respective acquisition executives' offices.

OT&E, on the other hand, falls under the purview of operational test activities (OTAs), established by law in each of the services. OTAs are organized under the chiefs of each service, not as part of the acquisition "chain of command" (Box 7.1). Congress directed this organization in order to ensure the independence of OT&E. The Director, OT&E (DOT&E), who reports directly to the Secretary of Defense and whose position also was established by Congress, provides independent oversight and reporting of OT&E.

LFT&E has the character of DT&E and so is conducted under the purview of the PM. However, Congress has vested oversight authority for LFT&E in the DOT&E.

Good management practice dictates that, early on in a program, the PM should begin forming teams among stakeholders from all relevant T&E agencies and offices. Potential team members could include representatives from contractors, user agencies, DT&E facilities, ranges, OTAs, and higher staffs with oversight authority. The PM's organization will likely include a team to integrate T&E activities into the overall acquisition strategy, as well as lower-level teams focused on specific test phases and events. Given the diverse cast of T&E players and stakeholders in most acquisition programs, good communication and coordination are essential for success.

BOX 7.1 IMPORTANCE OF OPERATIONAL TEST ACTIVITIES

In the 1990s, the Army and Marine Corps planned a joint program to acquire the 155-mm Lightweight Howitzer, a replacement for the M198 howitzer used by both services. The acquisition strategy entailed "shoot-offs" among prototype models from three firms. The winning firm would receive a full-scale development contract to provide howitzers for DT&E and OT&E. BAE Systems, a firm from the United Kingdom, won the competition, was awarded the development contract, and subsequently provided developmental model howitzers to the Army for testing. BAE planned to partner with a U.S. firm to manufacture the howitzers in the United States. However, initial DT&E by the Army on the developmental howitzers revealed several design flaws. The OTA—Marine Corps Operational Test & Evaluation Agency (MCOTEA)—subsequently determined that the developmental howitzers were not "production representative" and, thus were inappropriate for IOT&E because 1) the design flaws indicated design immaturity and 2) the developmental models were not produced using the same manufacturing processes, facilities, and workers that would produce items for fielding. To meet the "production representative criterion," a low-rate initial production phase was added to the program, slipping schedule by two years and increasing cost over $150 M [3].

7.5 PLANNING FOR T&E

Planning for T&E should begin as early as possible in an acquisition program's life cycle. This practice is consistent with good systems engineering, which calls for verification and validation planning at each stage of system development (see Fig. 7.1). Although T&E planners should develop long-range schedules and target dates for T&E events, they must also recognize that T&E, like acquisition itself, must be event driven rather than schedule driven. T&E should never be conducted simply to maintain adherence to program schedule, but rather it should occur according to and in support of the natural pace of system development.

7.5.1 ROLE OF PROGRAM DOCUMENTATION

Several program documents are key to proper T&E planning:

1) Capabilities, needs, and requirements documents describe the desired mission performance, give specific operational requirements to be attained, and state the intended operational environment and employment concepts for

the system under consideration. These will be essential in developing the following:

- *Critical technical parameters (CTPs)*: measurable characteristics that, when attained, allow for satisfaction of some operational requirement. An example of a CTP is engine thrust, which is necessary to satisfy a requirement to achieve a certain top speed. CTPs are most important in planning DT&E activities.
- *Critical operational issues (COIs)*: top-level operational effectiveness or suitability issues that are addressed in OT&E in order to determine whether a system can accomplish its mission. COIs are normally phrased as questions (e.g., "Does the system have sufficient mobility to operate in its intended environment?") and are usually decomposed into measures of effectiveness, performance, or suitability in OT&E planning. The extent to which these various measures are attained during OT&E indicates how COIs will be answered.

2) Threat assessments provide definitions of expected system threats and targets, which contributes to realism in T&E.

3) Specifications define requirements for system, subsystem, item, and component levels to be verified through T&E.

4) Acquisition strategies define overall business plans for acquisition programs. T&E planning must be reconciled with and integrated into acquisition strategies in order to properly balance cost, schedule, and performance objectives, and to provide timely and relevant information for decision makers.

7.5.2 T&E MASTER PLAN

Sound practice and current policy both require preparation, staffing, and approval of a T&E master plan (TEMP) to document the overall objectives, structure, and resource requirements for T&E activities in an acquisition program. The TEMP is required for entry into the full-scale development phase, but it remains a "living document" that must reflect changes to the program as they occur. As the road map for a program's T&E, the TEMP usually takes the form of a "capstone" summary document that provides a broad framework from which more detailed T&E events are planned and executed. The TEMP must be closely integrated with other program documents already mentioned, and its preparation should reflect a collaborative effort among T&E stakeholders.

Major sections of the TEMP include descriptions of the following: the system's intended mission, its key features, its threats, measures of effectiveness and suitability derived from requirements documents, and critical technical parameters; the T&E management structure and schedule of T&E activities; DT&E activities; OT&E activities, COIs, and LFT&E activities; and resources, including test articles, ranges and facilities, threat targets, operational forces support, modeling and simulation, funding, and training requirements.

7.5.3 T&E DESIGN

T&E activities are designed to provide necessary information to developers and decision makers at appropriate points in the project life cycle. T&E planners must therefore anticipate information needs early on in the program (as the requirements and acquisition strategy are being developed) and balance those with projected cost and schedule constraints.

A generic T&E process would consist of the following steps: 1) identification of information requirements at various points in the project life cycle; 2) analysis of the required information in order to determine its types and quantities, as well as the expected results of testing. (The nature of the required information will obviously vary according to the particular issue to be addressed.); 3) detailed planning for and collection of required data elements; 4) evaluation and synthesis of test data to produce needed information; and 5) use of that information in development and programmatic decisions.

7.5.4 PROJECTING SYSTEM PERFORMANCE — ROLE OF STATISTICS

Developers need methods to project with some confidence that their systems have acceptable probabilities of achieving desired performance levels. Statistical methods allow for inferences concerning performance to be drawn. Typically, developers and decision-makers desire high confidence and statistically significant results, which are achieved through larger test sample sizes. Of course, T&E is usually expensive and time consuming. Thus, T&E planners must strive for a balance between satisfying cost and schedule considerations and satisfying decision-makers' desires for reliable and precise performance estimates.

Some types of requirements lend themselves to verification through statistical methods, as in the case of a rifle test that requires numerous firings under controlled conditions to verify accuracy. Other types of requirements can be verified in a single test or demonstration, such as verifying that a truck fits into the cargo bay of an aircraft. Sample testing and statistical methods are employed largely in DT&E, when test conditions and data collection are more carefully controlled. The nature of OT&E, with its emphasis on operational realism, generally precludes statistically significant results.

7.5.5 MODELING AND SIMULATION

Use of modeling and simulation (M&S) to project performance has dramatically increased in recent years. The main advantage of M&S is that it allows for the repetition of many "trials" at relatively low cost (compared to actual T&E) and in ways that are neither unsafe nor environmentally harmful.

The major disadvantage is that M&S, by definition, entails abstractions of reality that cannot include many important "real-life" variables.

M&S is often used in conjunction with T&E to reduce costs. M&S can allow T&E planners to simulate test events and predict outcomes, which aids in validating test designs and assumptions. For certain types of systems, M&S provides information that cannot be obtained through T&E. For example, M&S is used almost exclusively in nuclear weapons testing and for missile defense system engagements. It also augments T&E by simulating certain nontestable events, such as two-sided weapons engagements and other unsafe scenarios.

Models and simulations can be assessed to determine their usefulness in certain T&E applications:

- Verification processes assess the extent to which the model or simulation functions according to its design. ("Does it work properly?")
- Validation processes assess the model's or simulation's fidelity in representing its intended segment of reality. ("Does it model the right thing?")
- Accreditation involves the acceptance of a model or simulation by an authoritative entity.

7.6 T&E IN THE PROJECT LIFE CYCLE

Chapter 1 described a project life cycle based on a set of critical decision points for acquisition programs. Figure 7.4 depicts T&E activities and emphases as they typically occur during this life cycle.

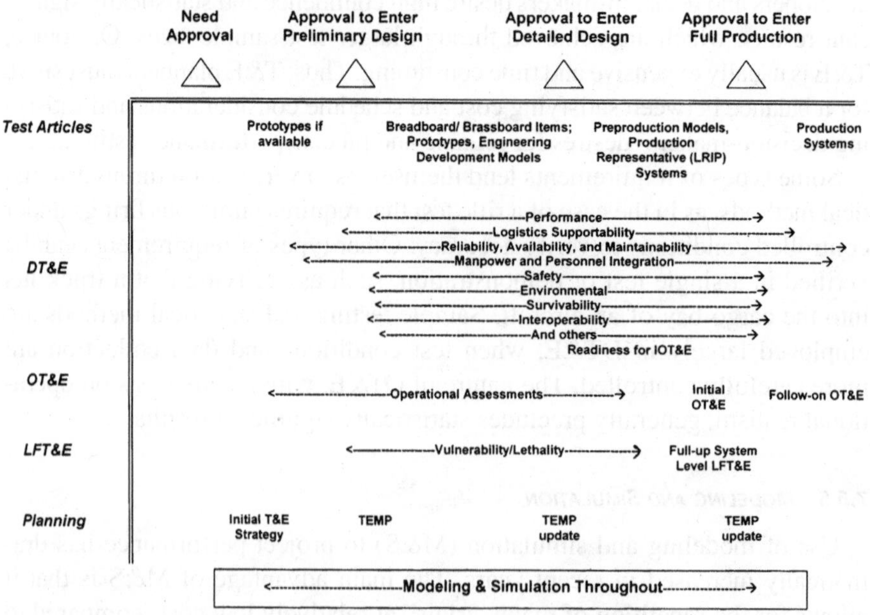

Fig. 7.4 T&E activities in the project life cycle.

7.7 CONCLUSION

Chapter 2 noted that the nature and extent of testing is one of the main areas of policy emphasis in defense acquisition. Engineers and decision makers will typically demand more T&E and higher-quality T&E in order to satisfy their desires for more and better information on a program's progress and risks. PMs, however, must balance those demands with cost and schedule constraints while ensuring compliance with applicable statutes. They should recognize that oversight agencies, OTAs, and others might scrutinize and criticize their T&E programs as inadequate. However, early engagement and collaboration with these parties in T&E planning can help allay such concerns. Thus, the best answer to the question, "How much T&E is enough?" is the answer that recognizes and accommodates the interests of all T&E stakeholders.

REFERENCES

[1] Forsberg, K., Mooz, H., and Cotterman, H., *Visualizing Project Management*, 3rd ed., Wiley, Hoboken, NJ, 2005, pp. 99–102.

[2] *Standard for Application and Management of the Systems Engineering Process*, IEEE Std 1220™-2005, Inst. of Electrical and Electronics Engineers, New York, 9 Sept. 2005, p. 12.

[3] Clark, P., *The XM777 Joint Lightweight 155 mm Howitzer Program (LW155): A Case Study in Program Management Considerations Concerning the Use of National Arsenal Assets*, MBA Professional Report, Naval Postgraduate School, Monterey, CA, 2003.

STUDY QUESTIONS

7.1 Describe the major purposes of T&E.

7.2 What are similarities and differences between DT&E, OT&E, and LFT&E?

7.3 What is the difference between operational effectiveness and operational suitability?

7.4 Describe three statutory requirements related to T&E.

7.5 Why is a teaming approach essential for effective T&E management?

7.6 Describe contributions of program documents in the development of the TEMP.

7.7 How do program documents contribute to development of the TEMP?

7.8 What is the importance of statistics in T&E? Of modeling and simulation?

7.9 Why are T&E plans and results so often criticized?

7.10 Watch the 1998 HBO film, *The Pentagon Wars* (starring Kelsey Grammer), on the test and evaluation phase of the M2 Bradley fighting vehicle. How much of the movie is "Hollywood," and how much of it rings true?

Chapter 8

RISK MANAGEMENT

RENE G. RENDON*
NAVAL POSTGRADUATE SCHOOL, MONTEREY, CALIFORNIA

LEARNING OBJECTIVES

- Define risk and risk management, and relate it to project management
- Understand the risk-management process
- Define risk planning, risk identification, risk analysis, risk-response planning, and risk monitoring and control, and explain the activities that occur in each
- Understand the risk-management tools and techniques used in project management
- Appreciate the value and use of a risk-management plan
- Be familiar with the current trends in risk management in the DoD project-management environment

8.1 INTRODUCTION

This chapter introduces the concept of risk and risk management and provides a discussion of the risk-management process. The chapter first provides an overview of risk-management fundamentals as they apply to the defense acquisition environment. It then discusses the risk-management process and the various activities involved in risk management. Risk planning, risk identification, risk analysis, risk-response planning, and risk monitoring and control will be specifically discussed as they apply to risk management in the defense acquisition environment. Finally, the chapter concludes with current trends in risk management in defense acquisition projects. In discussing risk management, references will be made to the *A Guide to the Project Management Body of Knowledge* (*PMBOK Guide*) [1], published by the

*Graduate School of Business and Public Policy.

Project Management Institute (PMI), as well as the *Risk Management Guide for DOD Acquisition*, published by the U.S. Department of Defense [2].

8.2 Risk Management in the Defense Acquisition Environment

Risk management in defense acquisition projects takes on an even more critical role than in traditional commercial projects. In traditional commercial projects, risk management is focused more on cost and schedule risks and less on technical risks. However, because of the mission of the Department of Defense and the technologies used in DoD weapons systems, risk management is a vital aspect of defense acquisition projects. With the increase in the DoD's weapon system technological complexity, high reliance on complex software, shortened acquisition timelines and the use of evolutionary acquisition approaches, and, of course, the inherent government funding instability, risk management will continue to be a critical aspect of project management in the defense acquisition environment.

In addition, the DoD's policy to use a knowledge-based approach in managing its acquisition projects has a direct impact on reducing technology, integration, and manufacturing risk. Box 8.1 illustrates the application of this knowledge-based approach to defense acquisition projects.

Project management and risk management go hand in hand. The project manager's job is to achieve the project's cost, schedule, and performance objectives. In trying to achieve these objectives, the project manager will be concerned with any risks involved as the project progresses. Thus, every decision the project manager makes is a reflection of the project's risk-management process.

Before we discuss the risk-management process, we will first define risk and risk management and relate risk management to the project life cycle.

Box 8.1 Knowledge-Based Approach

The GAO found that leading commercial firms pursue an approach based on knowledge, in which high levels of product knowledge are demonstrated at critical points in development. Programs take steps to gather knowledge that demonstrates that their technologies are mature; their designs are stable, and their production process are in control. This knowledge helps programs identify risks early and address them before they become problems. The result of a knowledge-based approach is a product delivered on time, within budget, and with the promised capabilities ([3], p. 11–12).

Risk is defined as "an uncertain event or condition that, if it occurs, has a positive or negative effect on the project's objectives" ([1], p. 373). The DoD defines risk as "a measure of future uncertainties in achieving program performance goals and objectives within defined cost, schedule, and performance constraints" [2]. As we can see, risk involves uncertainty in terms of events that will have an impact on achieving a project's cost, schedule, or performance objectives.

In addition, risk involves three elements: 1) a future event; 2) the probability of that future event occurring; and 3) the consequence of that future occurrence on the project's cost, schedule, and performance objectives.

These three risk elements can be conceptually defined to show that a project risk can be described as a function of the probability and impact of that risk event occurring, as follows [4]:

$$\text{Risk}_e = f\,(\text{probability, consequence})$$

These three elements serve as the basis for risk management and the processes used to manage risk.

Risk management can be defined as the act or practice of dealing with risk [4]. Risk-management activities should be applied through all phases of project management. In addition, risk-management processes should be integrated with other project-management processes, just as other functional processes and project activities are integrated. In the context of project management, specifically the project life cycle, the level of risk is highest in the beginning of the project and decreases as the project progresses to completion. As illustrated in Fig. 8.1, as the project progresses through the various

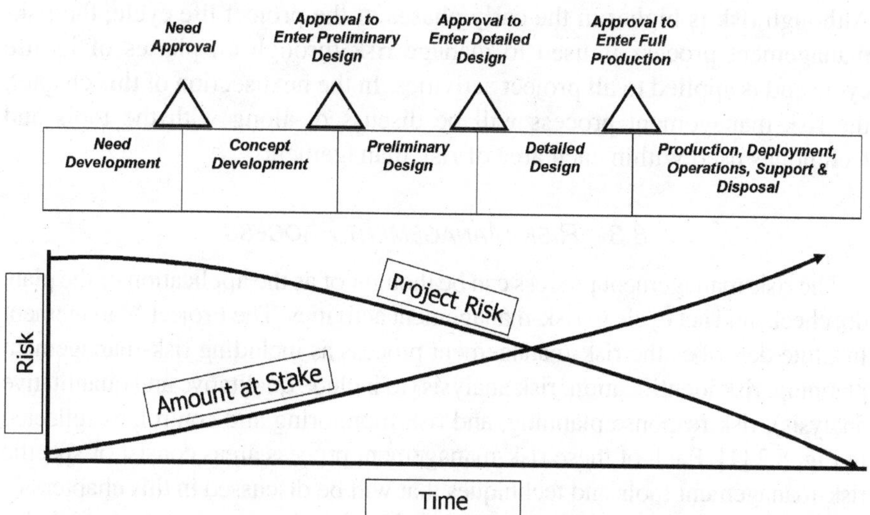

Fig. 8.1 Risk and project life cycle [4].

Box 8.2 Knowledge Acquisition to Mitigate Cost

"JTRS AMF has taken steps to develop knowledge prior to the start of system development. As part of the program's acquisition strategy, a presystem development phase started in September 2004 with the award of competitive system design contracts to two industry teams—led by Boeing and Lockheed Martin. Through this acquisition strategy, program officials expect competitive designs that will help mitigate costs and other risks" ([3], p. 91).

phases such as need development, concept development, preliminary design, detailed design, production, deployment, operations, support, and disposal additional project knowledge is gained. As this additional project knowledge is used in project activities, the level of uncertainty decreases, along with an associated decrease in project risk.

The relationship between risk and the project life cycle could obviously be seen in the acquisition of a new weapon system for the U.S. Department of Defense. As the acquisition is progressing through the need development, concept development, and preliminary design phases, the risk in achieving that weapon system's overall cost, schedule, and performance objectives will be higher than it will be when the weapon system is in the production, deployment, operations, support, and disposal phases. Box 8.2 illustrates how the acquisition strategy for the Joint Tactical Radio System (JTRS), Airborne, Maritime, Fixed-Station (AMF) will help to mitigate costs and other risks. Although risk is higher in the early phases of the project life cycle, the risk-management process is used to manage risk through all phases of its life cycle and is applied to all project activities. In the next section of this chapter, the risk-management process will be discussed, along with the tools and techniques used within each area of risk management.

8.3 Risk-Management Process

The risk-management process can be thought of as the application of the plan, do, check, and act cycle to risk-management activities. The Project Management Institute describes the risk-management process as including risk-management planning, risk identification, risk analysis (to include qualitative and quantitative analysis), risk response planning, and risk monitoring and control, as reflected in Fig. 8.2 [1]. Each of these risk-management process areas consist of specific risk-management tools and techniques that will be discussed in this chapter.

Although every project is different, and projects are used as the vehicle for acquiring weapons systems as well as services, the risk-management process

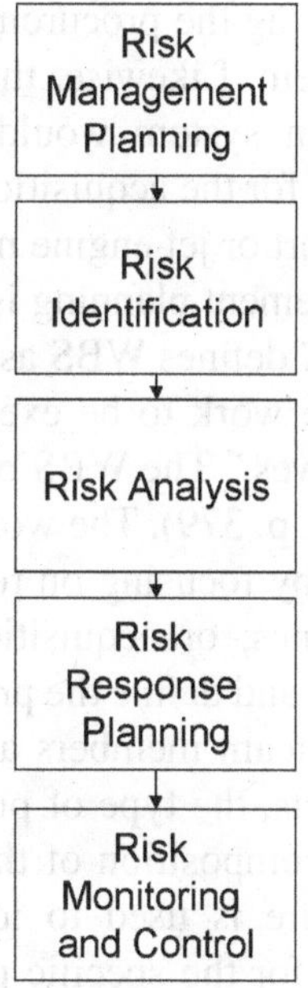

Fig. 8.2 Risk-management process ([1], p. 239).

is universally applied to all different types of projects. Indeed, both the project-management process and the risk-management process are universally applied to all types of projects. Although the process is the same, the difference is in how the program managers apply risk-management process activities to reflect the unique aspects of each project.

8.3.1 RISK-MANAGEMENT PLANNING

Risk-management planning is the process of deciding how to approach and conduct risk management for a specific project [1, p. 242]. Included in this planning is an analysis of the level, type, and the visibility of risk management that is needed for the specific project effort. The major tools used in risk-management planning include the project scope, work breakdown structure, and project risk breakdown structure.

The project scope defines the project objectives and describes the work that must be performed to deliver the product, service, or result within the project objectives [1]. The project scope encompasses all aspects of the project related to objectives, deliverables, cost and schedule estimates, and project stakeholders [1]. The specific aspects of the project scope will determine the level, type, and visibility of risk-management activities needed to ensure achievement of the project's costs schedule and performance objectives. Thus, a project involving the research, development, and production of a highly technical weapon system would require a different risk-management approach than a project involving the procurement of a commercial-off-the-shelf (COTS) product or system. Likewise, the risk-management approach needed for a defense weapon system would be different than the risk-management approach needed for the acquisition of critical defense services, such as aircraft logistics support or jet engine maintenance.

Also utilized in risk-management planning is the project work breakdown structure (WBS). The *PMBOK* defines WBS as "a deliverable-oriented, hierarchical decomposition of the work to be executed by the project team to accomplish a project's objectives." The WBS organizes and defines the total scope of the project effort ([1], p. 379). The work breakdown structure might decompose the project effort by focusing on requirements, processes, functional areas, technical baselines, or acquisition phases [2]. The WBS is employed not only to organize and define the project scope, but also to assign project tasks to these project team members and to develop project budget estimates. Just as in all projects, the type of project and scope of work will determine the level of the decomposition of the work breakdown structure. The work breakdown structure is used to identify the risk-management approach that will be needed for the specific project. The project scope, as reflected in the WBS, is used to develop the appropriate risk-management approach needed for a project.

Closely related to the work breakdown structure is the risk breakdown structure (RBS). The RBS graphically portrays the categories and subcategories within which risks can arise for a specific project. The risk breakdown structure parallels the work breakdown structure, as it also breaks down the work into different categories and subcategories [1]. A risk breakdown structure can have risk categories, such as technology, environment, organization, and requirement, as reflected in Fig. 8.3.

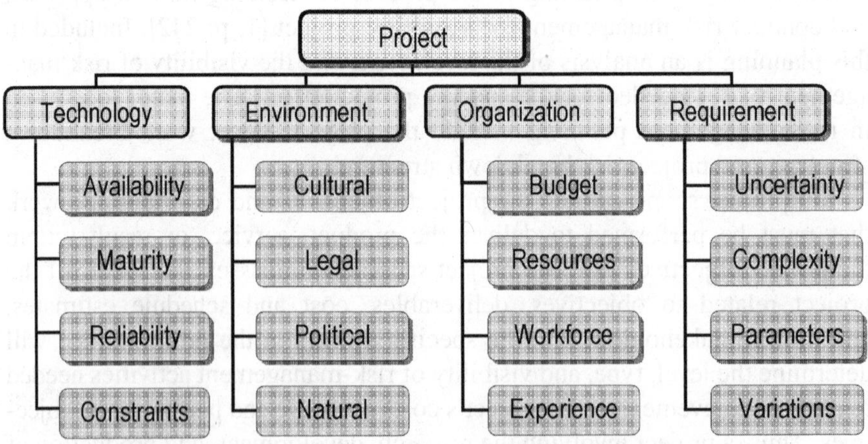

Fig. 8.3 Risk breakdown structure (adapted from [1], p. 244).

Through the use of the project scope, work breakdown structure, and risk breakdown structure, the project team can then determine how to approach and conduct risk management for a specific project. The result of risk-management planning, then, is a decision on what approach to use for risk management in terms of level, type, and visibility. Once the project team has conducted risk-management planning and has determined the risk-management approach for the specific project, it should develop a risk-management plan. The risk-management plan is a critical part of the overall project-management plan and documents the decisions made during the risk-management process. Thus, the risk-management plan will document the results of risk-management planning, as well as the other risk-management activities that follows.

Once the risk-management planning decision is made, the next phase of the risk-management process focuses on the identification of risks for the specific project.

8.3.2 *RISK IDENTIFICATION*

Risk identification is the process of determining which risks might affect the project and documenting those risk characteristics ([1], p. 246). Risk identification is performed as an iterative process throughout the entire project life cycle. As the project progresses through each phase, additional information that either increases or decreases the level of uncertainty in the project scope is obtained. As the project progresses and this new information is acquired, risk identification should be conducted continuously to identify new risks and to develop potential responses. Just as in risk-management planning, risk identification is based on information obtained from the project scope, work breakdown structure, risk breakdown structure, as well as project budget and project schedule.

Risk-identification activities include analyzing prior experience, brainstorming, learning lessons from similar projects, interviewing experienced subject-matter experts, analyzing the work breakdown structure, and analyzing key project processes or life-cycle phases. This would include key project processes such as the contracting process, the software development process, systems-engineering process, or any other process that plays a critical part of the project. Another key risk-identification activity is assumption analysis. Assumption analysis is basically determining how valid are the assumptions supporting the project effort. Assumption analysis explores the validity of the assumptions by analyzing their accuracy, consistency, or completeness ([1], p. 248). Another risk-identification technique includes checklist analysis, in which checklists are developed based on historical information and knowledge from previous projects, as well as other sources of information and diagramming techniques, where various project activities are diagrammed to identify root causes of risks ([1], p. 248). Box 8.3 illustrates how risk identification is being managed for the F-22 program.

Box 8.3 Risk Management and the F-22

"Risk management is considered an integral part of the F-22 E&MD Program and is an integrated part of each IPT's Iron Triangle responsibility. IPTs are charged with identifying their own risks as early as possible, determining the cause and significance, and developing and implementing effective risk-mitigation actions. Individual IPTs report on these actions to their next-higher-level IPT. Each IPT has procedures for resolving these risks itself or can refer to a higher tier for help" ([5], p. 39).

Once the potential project risks have been identified, the next step is to conduct an analysis of these risks.

8.3.3 Risk Analysis

Risk analysis is the process of determining the extent of the estimated risk by considering the probability of the risk-event occurrence and the possible impact or consequence of that risk event in terms of costs, schedule, and performance objectives of the project. It is important that risk analysis is conducted systematically using an approved, structured, repeatable methodology rather than a subjective approach [4]. This will ensure that risk-analysis results will yield accurate results. The *PMBOK* discusses two aspects of risk analysis: qualitative risk analysis and quantitative risk analysis ([1], pp. 249–260). Each of these aspects will be briefly discussed next.

Qualitative risk analysis is the initial analysis performed to identify the most important risk events in terms of likelihood of risk occurrence and impact of risk occurrence on project cost, schedule, and performance. The risk-identification phase of the risk-management process would have identified the estimated risks in the areas of cost, schedule, and performance project objectives. In the risk-analysis phase, the focus is on a more detailed analysis of the root causes of these identified risks. The DoD *Risk Management Guide* provides some questions to ask in assessing the root causes of cost, schedule, and performance risk events include the following:

1) Is there an impact to cost and at what level? Does the risk apply to acquisition costs or sustainment costs? How does the risk apply to life-cycle costs?

2) Is there an impact to schedule and to what level? Does this risk impact the critical path? Could the risk event change the project's critical path?

3) Is there an impact to technical performance and to what level? Does this risk impact operational, technical, or management performance areas (adapted from ([2], pp. 15–16)?

Although the preceding questions reflect a separate assessment for cost, schedule, and performance risk events, it should be stressed that these three areas are integrated within the project scope; risk events affecting one area will typically affect and impact (to some extent) the other two areas as well. Thus, a risk-event impacting schedule will most likely also impact cost and performance.

Qualitative risk analysis can be performed using a risk-analysis matrix. The risk matrix is a method used for classifying project risks based on the likelihood of the risk event occurring and the consequence of that risk event occurring. The steps in using the risk matrix involve the following:

1) Estimate the likelihood of the risk event occurring, considering the levels of likelihood and associated criteria.

2) Estimate the possible consequences or impact of the risk event, in terms of the projects cost, schedule, and performance objectives. This is done using the levels of consequence and associated criteria.

3) Determine the risk level for this risk event by plotting the estimated likelihood and consequence of the risk event on the risk matrix.

This analysis using the risk matrix results in the classification of the risk event as either low, medium, or high risk. The project team can then determine what additional risk analysis is needed for the higher-level-risk events, as well as take the appropriate risk response measures based on the level of risk for the risk event. The *Risk Management Guide for DoD Acquisition* provides an extensive discussion on using a risk matrix for conducting risk analysis. The following paragraphs summarize the use of the risk matrix, as discussed in the DoD guide.

As illustrated in Fig. 8.4, the risk matrix is a 5×5 matrix with the vertical axis reflecting the *likelihood* of the risk event occurring and the horizontal axis reflecting the *consequence* of the occurrence of the risk event. Both axes

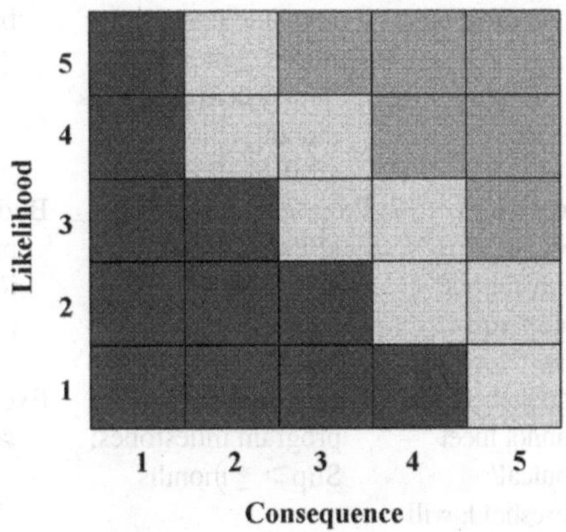

Fig. 8.4 Risk matrix (adapted from [2], p. 11).

TABLE 8.1 RISK-LIKELIHOOD LEVELS (ADAPTED FROM [2], p. 12)

Level	Likelihood	Probability of occurrence
1	Not likely	~10
2	Low likelihood	~30
3	Likely	~50
4	Highly likely	~70
5	Near certainty	~90

use a scale ranging from 1 to 5, with 1 being the lowest level and 5 being the highest level. Thus, the likelihood of the risk event occurring is based on a range of 1 (not likely, equating to a probability of occurrence of 10%) to 5 (near certainty, equating to a probability of occurrence of approximately 90%), as shown in Table 8.1.

Similarly, the levels and types of consequence or impact of the risk event ranges between level 1 (minimal or no consequence) to level 5 (severe impact) for each of the project's cost, schedule, and performance objectives. This is illustrated in Table 8.2.

TABLE 8.2 RISK-CONSEQUENCE LEVELS (ADAPTED FROM [2], p. 13)

Level	Technical performance	Schedule	Cost
1	Minimal or no consequence to technical performance	Minimal or no impact	Minimal or no impact
2	Minor reduction in technical performance for supportability, can be tolerated with little or no impact on program	Able to meet key dates; Slip < * month(s)	Budget increase or unit production cost increase; < ** (1% of budget)
3	Moderate reduction in technical performance or supportability with limited impact on program objectives	Miner schedude slip; able to meet key milestones with no schedule float; Slip < * month(s); Sub-system slip > * month(s) plus available float	Budget increase or unit production cost increase; < ** (5% of budget)
4	Significant degradation in technical performance or major shortfall in supportability, might jeopardize program success	Program critical path affected; Slip < * months	Budget increase or unit production cost increase; < ** (10% of budget)
5	Servere degradation in technical performance; cannot meet KPP or key technical/ supportability threshold; will jeopardize program success	Cannot meet key program milestones; Slip > * months	Exceeds APB threshold; > ** (10% of budget)

Furthermore, this risk analysis using the risk matrix can be applied at the different levels of the project work breakdown structure and can be tracked by WBS elements ([2], p. 14).

As previously stated, the *PMBOK* reflects two aspects of risk analysis: qualitative risk analysis and quantitative risk analysis. Once the qualitative analysis is performed using the risk-matrix tool, the project team can then apply the quantitative risk-analysis techniques to the risk events identified as significantly impacting the project's cost, schedule, and performance objectives ([1], pp. 254–260). Thus, the purpose of quantitative risk analysis is to numerically analyze the effect of the identified risks on project's cost, schedule, and performance objectives and to support the decision-making process during the management of the project ([1], p. 371).

Quantitative risk-analysis techniques include modeling and simulation such as the Monte Carlo technique, sensitivity analysis, expected monetary value analysis, and decision tree analysis. These techniques are used to quantify a project's estimated risk events and determine the risk events' probabilities, as well as to assess the probabilities of achieving a project's cost, schedule, and performance objectives [1]. A complete discussion of these quantitative techniques is beyond the scope of this chapter. However, there are a number of excellent references that provide extensive coverage of this topic. The point of this discussion is that quantitative risk analysis is a critical part of the risk-management process. Quantitative techniques would not be applied to all identified risks, only to the risks identified as having a significant impact on the project's cost, schedule, and performance objectives.

Finally, it should be stressed that a project's risk analysis is a "snapshot in time" ([2], p. 14). Just as a balance sheet reflects the financial condition of an organization at one point in time, the risk analysis of a project reflects the risk condition of that project at that specific point in time. As the project progresses, work is performed; time is consumed; costs are incurred; knowledge is acquired; uncertainty is reduced; and the risk condition of the project changes. Thus, risk analysis must be an iterative process, conducted continuously throughout the project's life cycle. Box 8.4 illustrates how risk analysis is being conducted for the Navy F/A-18E/F program.

Based on the results of the risk analysis, the project team can then use the risk-analysis results to determine responses and options for addressing these risks. This will be discussed next.

8.3.4 RISK-RESPONSE PLANNING

Risk-response planning is defined by the *PMBOK* as the "process of developing options and determining actions to enhance opportunities and reduce threats to a project's objectives" ([1], p. 260). This phase of the risk-management process is focused on determining the specific approach for

BOX 8.4 NAVY F/A-18E/F RISK MANAGEMENT

The Navy F/A-18E/F risk management program is managed by a risk-assessment board chartered by the Navy Program Manager. "The board meets quarterly, but reports monthly, in writing, on its activities. Membership consists of the Navy Program Office, NAVAIR matrix, DPROs, and the contractors. The board meets quarterly. A risk assessment of high, medium, or low, is based on five levels of uncertainty and five levels of consequences, and each risk item is assessed with this matrix" ([5], pp. 38–39).

addressing the identified risk events. The risk response will include the specifics of which approach to implement, when the approach should be implemented, who will be responsible for managing that specific risk, and the resources required to implement the risk response ([2], p. 18). The risk-response options include risk assumption, risk avoidance, risk mitigation, and risk transfer. These specific options will be discussed next.

Risk assumption is the conscious decision of acceptance of the risk event without making any changes to project scope or cost, schedule, and performance objectives ([4], p. 743). This strategy is selected when the level of the identified risk is not significant enough to warrant a change in the project's scope or objectives. Typically, the risk-assumption strategy includes the establishment of a contingency reserve (cost, schedule, and other resources) to handle actual risks that have been accepted ([1], p. 263).

Risk avoidance involves a change to the project scope—including changes to cost, schedule, and performance objectives—as a method of eliminating the root cause or the consequence of the risk event. This risk-response strategy is used when the level of the risk event is so high that the risk is detrimental to the achievement of the project's objectives. Risk-avoidance techniques can include changes in project concept, design changes, specification changes, or changes in the manufacturing process or procurement process ([4], p. 744).

Risk mitigation is a strategy that attempts to reduce the probability or likelihood of the risk event occurring, as well as the impact of that risk event. This risk-response strategy is not focused on eliminating the root cause or a consequence of the risk event, but merely on mitigating or reducing the level of the risk. Risk-mitigation techniques include creating alternative designs, increasing developmental test and evaluation or operational test and evaluation activities, using prototypes and engineering models, and using modeling and simulation in the product development. In addition, the use of control boards, such as configuration control boards and key-performance-parameter control boards, is an effective method of mitigating the risk of not achieving specific configuration or performance objectives ([4], p. 744).

Risk transfer is the process of shifting the risk impact to another party. The purpose of this strategy is to reallocate the risk, along with the responsibility for managing that risk, to another party. The risk-transfer strategy does not eliminate the risk, nor does it reduce the risk event's likelihood or consequence. Risk transfer merely gives responsibility for managing that risk to another party. This is typically arranged between the government and the prime contractor or between the prime contractor and its subcontractors. This risk-handling strategy typically involves paying a risk premium to the party assuming the risk to compensate the party for costs associated with the risk ([1], p. 262).

Typical risk-transfer techniques include financial instruments such as insurance, warranties, guarantees, bonds, and contracts [1]. Contracts are typically used for transferring risk from one party to another. For example, the use of cost-reimbursement contracts, in which all allowable, allocable, and reasonable costs are reimbursed by the buyer, transfers the cost risk from the seller to the buyer. On the other hand, fixed-price contracts, in which the seller is only paid a fixed price or adjustable price, transfer the cost risk from the buyer to the seller. Box 8.5 illustrates how cost-reimbursement contracts are used to transfer risk in the Joint Strike Fighter program. Figure 8.5 reflects the relationship between contract type and risk.

Once the risk-response strategies have been decided and implemented, the next phase of the risk-management process is monitoring and evaluating the effectiveness of the implemented risk-response strategies. This occurs during risk monitoring and control of risk management, which will be discussed next.

BOX 8.5 JOINT STRIKE FIGHTER AND RISK MANAGEMENT

"The GAO found that DoD is investing heavily in procuring Joint Strike Fighter aircraft before flight testing proves it will perform as expected. Producing aircraft before testing demonstrates the design is mature increases the likelihood of design changes that will lead to cost growth, schedule delays, and performance problems. Because the program will lack key design and testing knowledge, DoD plans to use cost-reimbursement contracts to procure early production aircraft. This type of contract places a substantially greater cost risk on the DoD and the taxpayers" ([6], highlights).

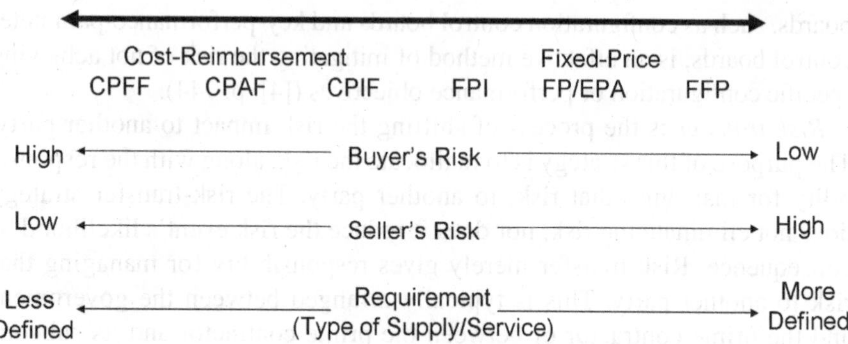

Fig. 8.5 Relationship between contract type, requirement, and risk.

8.3.5 Risk Monitoring and Control

Risk monitoring and control is a disciplined and systematic process of monitoring and evaluating the performance of the implemented risk-response strategies. The *PMBOK* describes risk monitoring and control as "the process of tracking identified risk, monitoring residual risk, identifying new risks, executing risk-response plans, and evaluating their effectiveness throughout the project life cycle" ([1], p. 373). Thus, risk monitoring and control can be thought of as a repeat of the prior processes of risk identification, risk analysis, and risk-response planning. As already stated, the risk-management process is an iterative process that occurs continuously throughout the project's life cycle. These risk-management activities are continuously performed during the risk monitoring and control process.

The *PMBOK* describes risk monitoring and control techniques to include the various project performance-management activities, such as variance and trend analysis and technical performance measurement. Risk monitoring and control includes using earned-value management analysis to identify project cost, schedule, and performance variances, as well as trends, and their effect on cost and schedule completion estimates. Risk monitoring and control also includes the measurement of various technical metrics (such as key cost and performance parameters) that would be part of a project's technical performance measurement. Using earned-value analysis and technical performance measurement allows the project team to compare actual project accomplishments with the planned cost, schedule, and performance achievements ([1], p. 266). Other risk monitoring and control techniques include program metrics of selected project-management processes, as well as schedule performance monitoring to assess how well the program is progressing ([4], p. 744).

During risk monitoring and control, the effectiveness of the implemented risk-response strategies is evaluated. If it is determined that these implemented strategies were not effective or did not accomplish their objective, then alternative risk-response strategies would be considered and implemented.

In addition, it might be determined that the implemented risk-response strategy was effective in lowering the level of risk for a specific risk event. In this case, the specific risk event would be reassessed and categorized with a lower priority. Box 8.6 illustrates how risks are monitored and controlled in the Littoral Combat Ship (LCS) program.

BOX 8.6 LITTORAL COMBAT SHIP AND RISK MANAGEMENT

The Navy stated that the LCS modular open system architecture strategy decouples core seaframe design and construction from the phased delivery of focused mission package payloads. A robust risk-management process tracks technologies under development to ensure they are matured and fulfill program requirements according to planned deployment timelines ([3], p. 102).

8.4 *RISK-MANAGEMENT DOCUMENTATION*

A final word on the risk-management process should focus on the importance of risk-management documentation. As already discussed in the risk-management planning phase, the risk-management plan is a critical part of the overall project-management plan, as it documents the decisions made during the risk-management process. The results of risk identification, risk analysis, to include the risk matrix, as well as other qualitative and quantitative analysis results, selected risk-response strategies, and risk monitoring and control activities should be documented in the risk-management plan. The risk-management plan should be a living and breathing document that is continuously updated as the project progresses through its life cycle. As risks are identified, risk-response strategies are implemented, and risk monitoring and control activities are performed, the risk-management plan should be updated to reflect the past decisions and current status of the project's risk profile. A well-documented and maintained risk-management plan serves as a critical audit trail of project-management and risk-management activities, but also serves as an excellent source to be used during a postproject audit for the identification of project best practices and lessons learned. The DoD *Risk Management Guide* provides the following sample format for a risk-management plan:

- Introduction
- Program summary
- Risk-management strategy and process
- Responsible executing organization
- Risk-management process and procedures

- Risk identification
- Risk analysis
- Risk-response planning
- Risk monitoring and controlling

8.5 CURRENT TRENDS IN RISK MANAGEMENT

Risk management continues to be a critical aspect of defense project management. As the DoD acquires even more advanced-technology weapon systems and crucial defense-related services in a resource-constrained environment, greater emphasis will be placed on meeting these program's cost, schedule, and performance objectives. The risk-management process plays a vital role in managing defense projects and achieving each program's objectives. With this increased emphasis on managing risks in resource-constrained projects, the following trends are occurring in the field of risk management:

Organizations are beginning to focus not only on the risks of not meeting project objectives, but also on the risks of not exploiting opportunities that might arise. Just as risk deals with the uncertainty of an event having a negative effect on projects objectives, opportunity deals with the uncertainty of an event having a positive effect on the project. In this case, opportunity is defined as "the measure of the probability of an opportunity event, a positive desired change occurring, and the desired impact of that event" ([7], p. 59). The risk-management process used to manage the threats against a project's objectives can also be used to manage the opportunities against the same objectives. The processes of planning, identification and analysis, response development, and monitoring and control should be used to address opportunities. The perspective of this risk-management process would focus more on the exploitation, enhancement, and sharing of opportunities than on the assumption, avoidance, mitigation, and transferring of risks ([1], p. 262). Some risk-management models include consideration of opportunities as well as risks. One example is the Contract Management Risk and Opportunity Assessment Tool (CROAT), which is designed to assist an organization in assessing the risk and opportunities associated with a pending potential or actual contract [8].

Another trend in risk management is the increased use of risk-management software tools. An example of this is the Technical Risk Identification and Mitigation System (TRIMS). TRIMS is a software tool that identifies areas of project risk, tracks program goals and responsibilities, and generates reports to meet the users' needs (data available online at http://www.bmpcoe.org/pmws/download/trims.html). TRIMS focuses more on technical risk than on cost and schedule. The philosophy here is that cost and schedule problems are simply indicators of more inherent technical issues. Additional information on TRIMS can be found at http://www.bmpcoe.org/pmws/download/trims.html.

Finally, another trend many organizations are adopting includes the use of tools for assessing organizational risk-management processes. The continuous process-improvement concept that evolved from the total quality management era of the 1980s resulted in organizations placing a greater emphasis on increasing the capability of organizational processes. Organizations have started using maturity models as a method for describing, measuring, and assessing organizational process capability. Some of these models include the Software Engineering Institute (SEI) Capability Maturity Model (CMM), the Project Management Institute's Organizational Project Management Maturity Model (OPM3), and the Contract Management Maturity Model (CMMM) [8]. Hillson proposes the Risk Maturity Model as a method for organizations to improve their approach to risk management, allowing them to assess their current level of maturity, identify realistic targets for improvement, and develop action plans for increasing their organization's risk capability [9]. The Risk Maturity Model provides four standard levels of maturity and outlines the activities necessary to move to the next higher level. The philosophy behind the Risk Maturity Model is that mature organizational risk-management processes are not the result of merely identifying risk-management techniques, providing risk-management training, and incorporating risk-management software tools. The basis of the Risk Maturity Model is that risk capability is a broad spectrum ranging from the lowest level of competence to the highest level of maturity. Only when organizations can assess their risk-management process maturity will they be able to improve their risk-management processes and strengthen their organizational competence in this critical project-management area [9].

8.6 CONCLUSION

This chapter discussed the risk-management process by focusing on the six different phases of managing risk. The chapter identified risk-management planning, risk identification, risk analysis, risk-response planning, and risk monitoring and control as discrete phases within the risk-management process. In the real world of defense-acquisition project management, these risk-management phases do not occur as separate and discrete activities. In reality, risk-management planning, identification, analysis, response planning, and monitoring and control activities overlap with each other and are integrated with other project-management processes, such as requirements development, cost estimating, quality management, and contract management. This chapter also discussed the importance of risk-management documentation and the value in developing and maintaining a risk-management plan. Finally, the chapter concluded with current trends in DoD risk management. Risk management will continue to be a critical aspect of DoD weapon systems acquisition. With the DoD's emphasis on evolutionary acquisition and its history of not following a knowledge-based approach, DoD weapon systems will continue to experience increase risks in meeting project cost, schedule, and performance objectives.

REFERENCES

[1] Project Management Inst., *A Guide to the Project Management Body of Knowledge (PMBOK Guide)*, 2004, pp. 237–268.

[2] Dept. of Defense, *Risk Management Guide for DoD Acquisition*, 6th ed., ver. 1.0, Washington, D.C., 2006, pp. 1–34.

[3] General Accounting Office, "Defense Acquisitions: Assessment of Selected Weapon Programs," GAO-07-406SP, Washington, D.C., 30 March, 2007.

[4] Kerzner, H., *Project Management: A Systems Approach to Planning, Scheduling, and Controlling*, Wiley, Hoboken, NJ, 2006, pp. 707–775.

[5] Johnson, R.V., and Birkler, J., *Three Programs and Ten Criteria*, RAND, Santa Monica, CA, 1996, p. 39.

[6] General Accounting Office, "Joint Strike Fighter: DoD Plans to Enter Production Before Testing Demonstrates Acceptable Performance," Washington, D.C., 15 March, 2006.

[7] Garrett, G. A., *Managing Complex Outsourced Projects*, CCH, Inc., Chicago, 2004, p. 59.

[8] Garrett, G. A., and Rendon, R. G., *Contract Management: Organizational Assessment Tools*, National Contract Management Association, McLean, VA, 2005, pp. 47–97.

[9] Hillson, D. A., "Towards a Risk Maturity Model," *The International Journal of Project and Business Risk Management*, Vol. 1, No. 1, Spring 1997, pp. 35–45.

STUDY QUESTIONS

8.1 Why is risk management more important in defense acquisition projects?

8.2 Define risk and risk management.

8.3 What are the three elements of risk?

8.4 How does risk relate to the project life cycle?

8.5 Define and discuss risk-management planning.

8.6 Describe the use of the work breakdown structure and the risk breakdown structure in risk management.

8.7 Define and discuss risk identification.

8.8 What are some risk identification techniques?

8.9 Define and discuss risk analysis.

8.10 Describe the use of a risk matrix in risk management.

8.11 Define and discuss risk-response planning.

8.12 What are four risk-response options?

8.13 Describe the relationship between contract type, requirement, and risk?

8.14 Define and discuss risk monitoring and control.

8.15 Discuss how earned value management can support the risk monitoring and control process.

8.16 Discuss the value of developing and maintaining a risk-management plan.

8.17 What are some current trends in risk management?

Chapter 9

Manufacturing and Quality

Michael Boudreau*
Naval Postgraduate School, Monterey, California

Learning Objectives

- Describe the defense industrial base
- Describe the defense manufacturing spectrum
- Identify manufacturing activities and considerations at various points during the project life cycle
- Understand the connection between manufacturing and systems engineering
- Describe manufacturing readiness levels
- Describe producibility considerations as they pertain to manufacturing efficiency
- Identify the six manufacturing elements
- Describe manufacturing initiatives such as lean and enterprise resource planning (ERP)
- Describe the "house of quality" construct
- Describe quality management systems and quality initiatives
- Outline DoD policies on quality

9.1 *Introduction*

This chapter will examine manufacturing and quality from several different perspectives. First will be a review of the manufacturing spectrum because manufacture of bullets, trucks, airplanes, or ships is vastly different by almost any measure, from complexity to quantity, to cost, to the required manufacturing time. Manufacturing planning and preparations stretch across the preacquisition and acquisition phases, addressing the manifold issues related to production capability, capacity, affordability, and efficiency. Discussion of

---

*Senior Lecturer, Graduate School of Business and Public Policy.

manufacturing will be linked to the systems-engineering-process (SEP) because the manufacture of warfighting implements is the culmination of a developmental effort, without which manufacturing is not possible.

Quality will also be examined through multiple lenses. Quality considerations also span all acquisition phases, addressing identification of critical processes and metrics, test and measurement equipment, process controls, and quality management systems. Particular attention will be paid to quality systems, such as ISO 9001, which provide complete quality management frameworks that can be used to guide manufacturing processes and ultimately furnish products that satisfy the needs of military users.

Manufacturing and quality experts must be members of Integrated Product Teams (IPTs) if they are to insure that manufacturing and quality areas of interest are addressed as part of the multi-functional exchange occurring throughout product development.

9.2 Defense Industrial Capability

The industrial base, which accomplishes the design and manufacture of U.S. warfighting systems, includes thousands of different facilities—including government and contractor laboratories and design facilities; vendors that produce components, modules, and pieceparts; large contractor production plants, government arsenals and overhaul facilities; and extensive Government and private-sector test facilities. All of these activities comprise the U.S. defense capability or the defense industrial base. Manufacture of DoD warfighting systems depends on prime contractors and subcontractors, subtier producers, and vendors. Current policy is to use commercial or nondevelopmental items or content where feasible in order to save resources and obtain access to innovative technologies developed in the private sector.

Although most warfighting systems are built in the United States, virtually all systems have varying degrees of foreign content—from base metals and rubber products to very sophisticated components. DoD performs comparative testing of foreign materiel and occasionally acquires foreign warfighting systems, such as Fox NBC reconnaissance vehicles and AV-8 Harrier aircraft.

9.3 The Manufacturing Spectrum

Defense contractors produce a multitude of different warfighting systems, including ships, aircraft, missiles, tanks, trucks, communication systems, and munitions. These items come from many different commodity areas, and, quite obviously, the manufacturing processes look very different from one commodity area to another. Part of the difference is based on the numbers of items produced. In a simple way, defense manufacturing can be categorized by numbers of product, continuity of the manufacturing process, and the

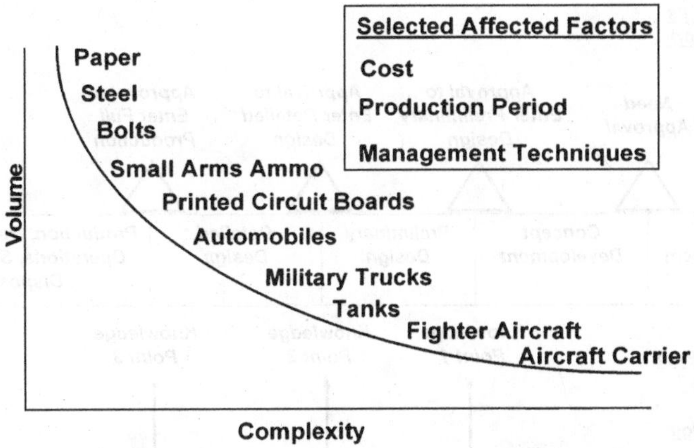

Fig. 9.1 Defense manufacturing spectrum.

duration of manufacturing (Fig. 9.1). For example, large ships are typically designed and constructed singly and might take more than five years to construct. Ship production quantities might range from a single ship to several dozen. Military aircraft might be produced in greater quantity than ships, with the manufacturing period of an individual fighter/attack aircraft taking about two years. Tanks and trucks are likely produced in larger quantities, perhaps in the thousands, than military aircraft. Although tanks might be assembled in several weeks, some of their more complicated components, such as the engine or a fire control device, might take an even longer time to produce. Moving down the complexity spectrum, military trucks might be built in quantities of tens of thousands, with individual trucks being built within a few days.

Some of the constituent parts of warfighting systems are manufactured in bulk on continuous production runs. For example, steel plate used in ships or tanks is rolled in large production runs that are continuous or nearly continuous. Small-arms ammunition is produced at vastly higher rates than are weapon systems. The cost and elapsed time for assembly of a round of ammunition are miniscule, and (even acknowledging the safety aspects) there is much less management involvement per round of ammunition than is involved in the assembly of a major weapon system. The point is that all of the different items shown in the manufacturing spectrum figure are "manufactured," but the production environments might be very different indeed.

9.4 MANUFACTURING CONNECTION TO THE PROJECT LIFE CYCLE AND SYSTEMS ENGINEERING

Chapter 1 introduced the phases and milestones of the acquisition project life cycle, and it also related the life cycle to "knowledge points" proposed by

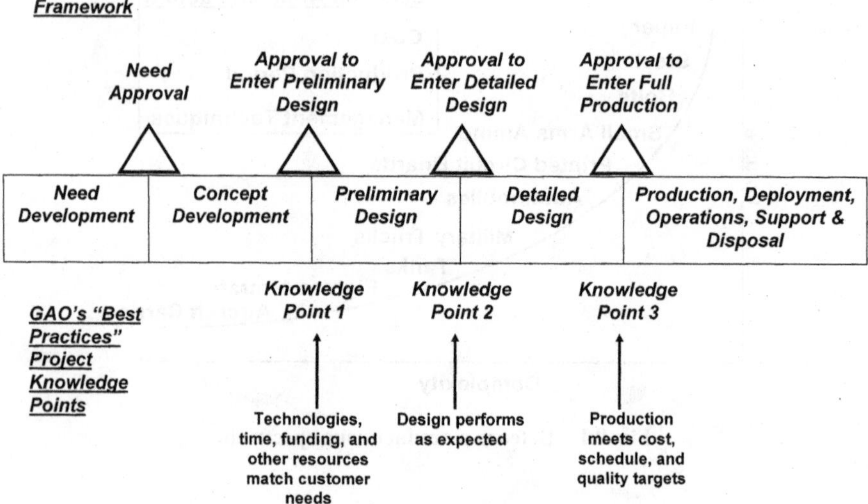

Fig. 9.2 Project life cycle.

the Government Accountability Office (GAO). This section discusses manufacturing activities that are relevant to this life cycle and these knowledge points (see Fig. 9.2). It also discusses relevant aspects of systems engineering as they relate to manufacturing activities.

9.4.1 CONCEPT DEVELOPMENT

During this preacquisition phase, the sponsor or "institutional user" has the lead, but the program-management office (PMO) and contractors are also involved in the work effort. Management attention is focused on the ability of manufacturing to produce the eventual warfighting system. Although the design of the warfighting system is not "nailed down," developers are at work with users and contractors to identify likely technologies that will be integrated into the evolving system. For example, if the new system will need to be stealthy, incorporate high-energy weapons, and use advanced "fire and forget" fire-control systems, then the question must be asked: Are the anticipated manufacturing technologies mature enough to ensure stable production at affordable cost, not having to resort to high-risk exotic techniques under laboratory conditions?

The phase of concept development also provides useful information about technology maturity, including insights into production facilities and processes that will be required to produce systems or components that have not been manufactured before. This phase provides the opportunity for project management to ensure that the answers are known to the kinds of unique issues raised in the preceding paragraph.

9.4.2 KNOWLEDGE POINT 1

Knowledge point (KP) 1 occurs when a match is made between the customer's needs and the available resources: technology, design, time, and funding. To achieve this match, the technologies necessary for meeting essential product requirements must be demonstrated to work in their intended environments. In addition, the product developer must complete a preliminary product design using systems engineering to balance customer desires with available resources. KP 1 affects manufacturing because the maturity of required technologies includes manufacturing technology readiness. Additionally, improved knowledge of the funding required to produce a system also results from KP 1.

9.4.3 PRELIMINARY AND DETAILED DESIGN

During the design phases, manufacturing engineers and technicians work with design teams to ensure adequate attention is given to designing for producibility. Additionally, manufacturing and quality practitioners are particularly focused on the detailed planning for facilities and processes that will support production of warfighting systems, components, assemblies, and parts. These must be created in an efficient manner that controls product variation. Are design drawings detailed with specific information necessary to successfully manufacture and check key components? Is factory equipment being selected that is consistent with the design requirements and required production rates? Is test equipment being put in place that can verify the attributes that are critical to quality?

9.4.4 KNOWLEDGE POINT 2

Knowledge point 2 occurs when developers demonstrate that the product design meets performance requirements. Program officials are confident that the design is stable and will perform acceptably when at least 90% of engineering drawings are complete. Engineering drawings reflect the results of testing and simulation and describe how the product should be built. KP 2 smoothes transition into production by better ensuring that the system design has been completed and will support the production process.

If the program strategy includes *low-rate initial production,* processes must be verified, and product variation must be shown to be within acceptable bounds. Are tooling, processes, and skilled personnel in place to control manufacturing reliably within the required tolerances? Can warfighting systems that perform as required, at an affordable production cost, and at an acceptable production rate be manufactured?

9.4.5 KNOWLEDGE POINT 3

Knowledge point 3 occurs when the product is reliable and can be manufactured within cost, schedule, and quality targets. An important indicator of this is when critical manufacturing processes are in control and consistently producing items within quality standards and tolerances. Another indicator is when a product's reliability is demonstrated through iterative testing that identifies and corrects design problems. KP 3 encourages careful selection, validation, and control of processes to ensure that they provide the necessary product quality. Process selection and control is further explained in the quality section of this chapter.

9.4.6 FULL-RATE PRODUCTION AND DEPLOYMENT PHASE

As manufacturing ramps up to full production, the product developers must emphasize control of product variation through statistical process control, six sigma, and other techniques; at the same time, they must pursue cost efficiencies through *lean* or other techniques that promote efficiency and reduce wasted effort.

9.4.7 BASELINES AND SPECIFICATIONS

Baselines and specifications (Table 9.1) guide and inform manufacturing activity throughout the developmental process—first by the system performance specification and functional baseline, which describe performance attributes that the warfighting system must be able to achieve. As the systems-engineering process produces more refined definition of components and subsystems, this information is documented in the functional performance specification and allocated baseline. At such time as functional (design-to) specifications have been assembled to support detailed design, they are documented by the preliminary design review (PDR). The detailed design effort includes drawings and other documentation required for system production. The completed technical data are reviewed during a critical design review (CDR); the end product of which is the initial product baseline. These documents and baselines form essential underpinnings for manufacturing and quality processes.

TABLE 9.1 SPECIFICATIONS AND BASELINES

Specifications	Baselines
System performance	Functional
Item performance (design to)	Allocated
Item detail (build to)	Product

9.4.8 REVIEWS AND AUDITS

Several of the system-engineering reviews are directly related to manufacturing and quality.

- Production readiness reviews (PRR) are focused on such issues as the readiness of the technical data, the adequacy of manufacturing facilities, selection and acquisition of suitable tooling and other required equipment, selection of suppliers, validation of manufacturing and quality processes,

> **BOX 9.1 PROGRAM OUTCOMES RELATED TO PRODUCTION READINESS ([1], P. 8)**
>
> DoD programs that had more successful outcomes used key best practices to a greater degree than others. For example, the AIM-9X missile program completed 95% of its engineering drawings at the critical design review because it made extensive use of prototype testing (to demonstrate the design met requirements) coupled with design reviews that included program stakeholders. The F/A-18-E/F program eliminated over 40% of the parts used to build predecessor aircraft to make the design more robust for manufacturing. It also identified critical manufacturing processes, bringing them under control before the start of production. Both programs developed products that evolved from existing versions, making the design challenge more manageable.
>
> On the other hand, DoD programs with less successful outcomes did not apply best practices to a great extent. At their initial manufacturing decision reviews, the F-22, PAC-3, and ATIRCM/CMWS had less than one-third of their engineering drawings—in part because they did not use prototypes to demonstrate the design met requirements before starting initial manufacturing. On the F-22 program, it was almost three years after this review before 90% of the drawings needed to build the F-22 were completed. Likewise, at their production decision reviews these programs did not capture manufacturing and product reliability knowledge consistent with best practices. For example, the PAC-3 missile program had less than 40% of its processes in control. As a result, the missile seekers had to be built, tested, and reworked an average of four times before they were acceptable. The F-22 entered production despite being substantially behind its plan to achieve reliability goals. As a result, the F-22 is requiring significantly more maintenance actions than planned [2].

and adequacy of quality and quantities of purchased and manufactured components (see Box 9.1).

- Physical configuration audits (PCA) compare dimensions and other physical characteristics of manufactured items to the technical data to ensure that they match and are correct.
- Functional configuration audits (FCA) verify that performance characteristics conform to specification requirements.

9.4.9 Configuration Management

Configuration management (CM) crosses numerous functional disciplines and, although not unique to manufacturing, is essential. CM provides the processes by which configuration is controlled and ensured. To ensure configuration and synchronize required configuration changes, CM must connect to every part of a manufacturing organization and also reach outward to suppliers and customers. Configuration changes are reflected by changes to part numbers, signaling to people throughout a facility that a configuration change has been made. These notifications of a configuration change must be communicated to quality inspectors, machinists, assemblers, and the supply warehouse. Configuration changes might require changes to test equipment and various manufacturing tools and processes. Configuration-change documentation must specify when the change will become effective and the serial number of items affected and will result in revisions to other documentation (such as the manufacturing operations instructions, quality check sheets, cost accounting records, etc.). Although this change process seems very tedious, disciplined CM processes are necessary to ensure that the correct product is being manufactured and that the manufacturing organization, suppliers, and customers all have a synchronized, common understanding of the configuration of the product.

9.4.10 Risk Management

Risk management is essential activity in manufacturing because of uncertainty and the potentially severe consequences that might result. In the 1980s, the Defense Science Board commissioned a study on risk management. An output of this effort was DoD 4245.7-M, "Transition from Development to Production," published in February 1985 [3]. This publication is frequently referred to as "Willoughby's Templates," named after W. J. Willoughby, Chairman of the 1982 Defense Science Board Taskforce on "Transition from Development to Production." Willoughby's template approach to risk management recognized that many different programs in the past had made similar errors during their transition to production. The templates comprised categories such as funding, design, test, production, the transition plan, facilities, logistics, and management. Specific production templates included areas such as manufacturing planning, quality manufacturing process, piece-part control, subcontractor control, and defect control. Each of the preceding templates contained a set of generic questions to guide risk-management planning for a specific process. The DoD offers an automated risk-management program, called the Technical Risk Identification and Mitigation System (TRIMS), that incorporates and expands Willoughby's templates. This program provides standard question sets that can be modified for specific applications. TRIMS has expanded beyond Willoughby's original categories to address software systems.

9.5 MANUFACTURING READINESS LEVELS

DoD has developed manufacturing readiness levels (MRLs) [2] as measures for assessing manufacturing maturity. MRLs provide decision makers with an understanding of the relative maturity and risks associated with a project's manufacturing technologies, products, and processes. Manufacturing maturity considerations include tooling, processes, and techniques that permit reliable, stable production. MRLs address nine "threads" that should be addressed in all phases of product development: technology and industrial base, design, materials, cost and funding, process capability and control, quality management, personnel, facilities, and manufacturing planning, scheduling, and control.

9.6 PRODUCIBILITY

Producibility engineering is mentioned several places in this chapter and is an important design activity. Accomplished in an IPT environment, producibility engineering can help achieve a balance that includes ease of manufacturing, ease of logistics, and reduction in total ownership cost (R-TOC). New designs are not automatically "manufacturable," suitable, effective, reliable, maintainable, and affordable. However, through multifunctional dialogue, such as through the employment of IPTs, a balanced design is possible and ought to be mandatory. Boothroyd Dewhurst's "Design for Manufacture and Assembly" [4] is an automated producibility program that can be employed to make designs more efficient to manufacture. Producibility engineering is particularly useful during the design stage of a program and is aimed at getting the solution right the first time.

Conversely, much of the emphasis of value engineering is focused on correcting design or manufacturing that was not planned appropriately at the outset. It always makes sense to try to achieve a perfect solution the first time, but, of course, such effort is never 100% successful. Any solution can be improved. Indeed, that belief is the thrust of lean techniques, which continually strive for improvement throughout the lifetime of the endeavor.

9.7 MANUFACTURING ELEMENTS

Six manufacturing elements (the 6 Ms) are very useful in planning manufacturing activities, in that they provide a framework for identifying potential problem areas. The 6 Ms are often depicted in *cause and effect diagrams* (commonly referred to as *Ishikawa* or *fishbone* diagrams), which are widely used by quality and reliability engineers to guide root-cause analysis of failures. The 6 Ms are shown in Fig. 9.3.

9.8 MANUFACTURING TOOLS AND TECHNIQUES

The tools and techniques covered in this section are also applicable to the subject of quality, covered in more detail in the following.

9.8.1 LEAN (JUST IN TIME)

Lean methods have their roots in the Toyota Production System (see Sec. 9.10.5). However, the notion of lean has been distilled and described in *Lean Thinking: Banish Waste and Create Wealth in Your Corporation* [5], which describes five process steps:

- *Value*: value is defined by the customer (i.e., required attributes, price, and delivery time).
- *Identify the value stream*: manufacturers must put together all of the steps necessary to provide a product that the customer values.
- *Flow*: work-in-process should move through its process continuously, without stops, delays, or backflow.
- *Pull*: the customer should *pull* (demand) the product, rather than the producer *pushing* the product to the customer.
- Perfection: perfection is a never-attained goal but keeps developers continually focused on improving the process.

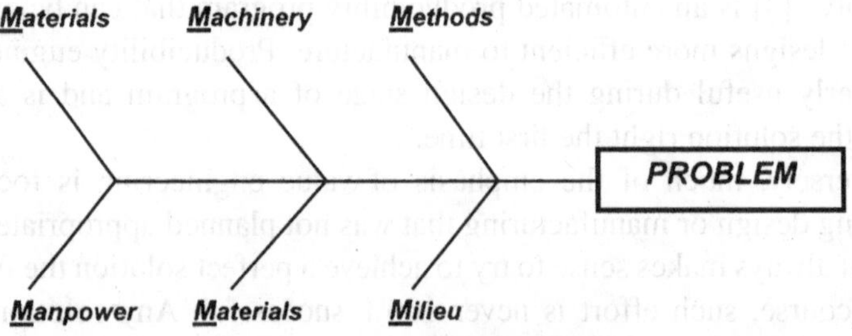

Fig. 9.3 Six manufacturing elements.

9.8.2 ENTERPRISE RESOURCE PLANNING

Enterprise resource planning (ERP) is an automated information system used by companies and organizations to connect the necessary databases that facilitate sharing of information and knowledge across the enterprise. Since the 1970s, automated information systems—material requirements planning (MRP)—have been used in manufacturing to purchase and control inventory in support of production. Such automated systems could be used to time sequence the order of materials so that inventory would arrive just before it was needed on the manufacturing floor. Later, in the 1980s, algorithms were added to MRP to assist in maximizing manufacturing flow (MRP II). In the 1990s, ERP evolved as systems that supported most or all of the organization. SAP R/3, an illustrative ERP system in wide use today, includes application modules for activities such as sales and distribution, materials management, production planning, financial accounting, and human resources.

9.8.3 COMPUTER-AIDED DESIGN (CAD), COMPUTER-AIDED MANUFACTURING (CAM), AND COMPUTER-AIDED ENGINEERING (CAE)

The preceding are all productivity enhancements that support design, engineering, and manufacturing functions.

- CAD has taken design off the drafting board and made it electronic. Not only can design be accomplished more efficiently and be changed more quickly on design workstations, it can also be coordinated much more rapidly in a multifunctional environment. Design drawings can be produced as three-dimensional solid models that can be shown to customers to obtain some early confirmation of the design. Three-dimensional solid models can be promptly turned into illustrations in technical manuals.
- CAM can use the digital design to program computer numerically controlled (CNC) milling machines and lathes to automatically control machining of required parts. CAM can be linked to automated dimensional inspections that are accomplished on coordinated measuring machines. Digital designs can be linked with automated purchasing systems that can order internal parts and components or entire assemblies or end items.
- CAE is used to analyze emerging designs in different ways. For example, the design of a printed circuit board can be analyzed for heat buildup through CAE. A ship design can be evaluated for structural characteristics through finite element analysis. Intricate mechanical designs can be reviewed for appropriate interferences between adjacent parts or assemblies (i.e., parts that do not fit together properly). System designs can be operated in virtual reality; for example, digital vehicular designs can be operated across digital terrain via the NATO reference mobility model.

The beauty of CAD/CAM/CAE is that information can be input once and then used many different times and in different ways by individuals across the organization and, via the Internet, even by suppliers, customers, and maintainers.

9.8.4 CAPABILITY INDEX C_{PK}

C_{PK} is obtained by comparing design tolerance (the requirement) against process capability (Table 9.2). For example, if the design tolerance of a particular dimension must be within plus or minus 1 in., and the process capability can be controlled within plus or minus 0.5 in., then C_{PK} would equal 2.0 (i.e., 1 divided by 0.5 = 2), meaning that the manufacturing *defect rate* should be very, very small.

9.9 QUALITY MANAGEMENT

Quality begins with the perspective of the customer—that is, the aspects that are important to the customer—in terms of performance of a specific capability, within a price (and probably, delivered by a required date). Performance can include many different attributes, such as speed, accuracy, weight, ease of operation, ease of maintenance. Some characteristics, such as speed, accuracy, and weight, can be measured easily and directly. Other desired attributes, such as reliability and availability, also are measurable, but might not be as directly as the aforementioned. Other measures, such as ease of operation, might be more difficult to measure directly and might be qualitative, rather than quantitative. Some aspects of appearance, such as surface coatings and finish, might be partially measurable and partially qualitative. All of the foregoing can be listed as customer-specified requirements, prioritized, and translated into quality management considerations through various techniques, such as quality function deployment, sometimes called house of quality (Fig. 9.4). Some customer performance requirements might be in opposition to one another and, thus, require balance through an agreeable compromise.

Whatever capabilities the customer identifies as most important can be managed and controlled through the manufacturing process, so that the end

TABLE 9.2 C_{PK} INDEX AND PROBABILITY OF A DEFECTIVE PART ([1], P. 39)

Manufacturing process capability C_{pk}	Associated defect rate
C_{pk}—0.67 (not capable)	1 in 22 parts produced[a]
C_{pk}—1.0 (marginally capable)	1 in 370 parts produced[a]
C_{pk}—1.33 (industry standard)	1 in 15,152 parts produced[a]
C_{pk}—2.0 (industry growth goal)	1 in 500,000,000 parts produced[a]

[a]*Defect rates*, as shown above, are an alternative way of expressing *defects per million opportunities*.

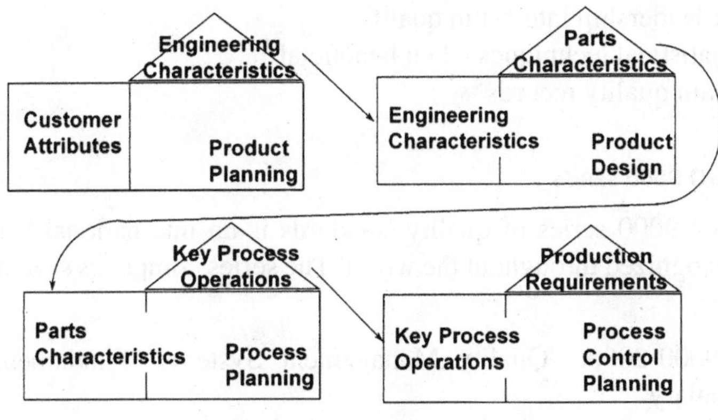

Fig. 9.4 House of quality.

product reflects those aspects the customer most desires. Much of the responsibility of quality management is to identify and understand those aspects that are most important to the customer and help to identify and control the processes and conditions that influence the attributes which reflect quality. This monitoring is often attempted by *inspecting quality in*, that is, inspecting and rejecting the product that does not conform in those areas that are critical to quality. A more efficient approach is to control the processes that influence attributes that are critical to quality (CTQ). Quality personnel help to ensure product quality through audit of processes and product. But the optimal situation would be for manufacturing personnel to take control of the processes that affect CTQ attributes. This is called *building quality in*. Note the difference between building quality in and inspecting quality in. Normally, building quality in is much more efficient and costs less than inspecting quality in.

9.10 QUALITY-MANAGEMENT SYSTEMS

In support of manufacturing, quality managers must identify the processes that create or trace the measurable CTQ attributes the customer has specified. Any quality-management system must be able to control the processes that produce CTQ attributes, audit the processes and the product, and improve both. Some illustrative quality functions, needed by virtually any quality-management system are as follows:

- Control key processes that are critical to quality
- Control vendor-supplied material
- Conduct root-cause analysis and perform corrective action to correct defects found during manufacturing or in items "in the field"
- Segregate and dispose of discrepant material
- Enable continuous process improvement

- Foster leadership interest in quality
- Use statistical techniques when beneficial
- Maintain quality records

9.10.1 ISO 9000 SERIES

The ISO 9000 series of quality standards is an international benchmark widely recognized throughout the world. The series comprises several related principles:

- ISO 9000-2000, "Quality Management Systems—Fundamentals and Vocabulary"
- ISO 9001-2000, "Quality Management Systems—Requirements" (Fig. 9.5)
- ISO 9004-2000, "Quality Management Systems—Guidelines for Performance Improvements"

9.10.2 TOTAL QUALITY MANAGEMENT AND TOTAL QUALITY LEADERSHIP (TQM/TQL)

TQM/TQL became very popular during the 1980s and into the 1990s but have fallen out of favor, probably because they were applied in organizations that were ill suited for their approaches. Elements of total quality include customer focus, obsession with quality, scientific approach, long-term commitment, multifunctional teamwork, continual process improvement, education and

Quality-Management System - Requirements

1 Scope

2 Normative reference

3 Terms and definitions

4 Quality management system

- Documentation requirements

5 Management responsibility

- Management commitment
- Customer focus
- Quality policy
- Responsibility, authority and communication
- Management review

6 Resource management

- Provision of resources
- Human resources
- Infrastructure
- Work environment

7 Product realization

- Planning of product realization
- Customer-related processes
- Design and development
- Purchasing
- Production and service provision
- Control of monitoring and measuring devices

8 Measurement, analysis and improvement

- Monitoring and measurement
- Control of nonconforming product
- Analysis of data
- Improvement

Fig. 9.5 ISO 9001-2000.

training, freedom through control, unity of purpose, and employee involvement and empowerment [6]. These elements are valuable and continue to be useful, but they do not form a complete quality-management system. These elements could be applied to different quality-management systems; indeed, they show up in other initiatives, such as Lean Six Sigma.

9.10.3 STATISTICAL PROCESS CONTROL

Statistical Process Control (SPC) was developed by W. A. Shewhart, a statistician who experimented, beginning about 1925, with the control of product variation through the use of statistical techniques. Product variation is plotted on control charts (Fig. 9.6) that display changes in variation over time. Typically, statistics such as *average* and *dispersion* are plotted for each sample. Control charts show trends (i.e., changes over time) which indicate that something has gone amiss with the process and should be fixed. The goal of SPC is to find and fix any problems that manifest themselves in trends before they result in the manufacture of defective product.

9.10.4 SIX SIGMA (6σ)

The Six-Sigma approach began at Motorola in 1979 when a corporate executive, Art Sundry, stated in a meeting that, "The real problem at Motorola is our quality stinks" [7]. This pronouncement led to a change in the current manufacturing thinking and a new understanding that improving quality could actually reduce costs. This was accomplished through disciplined

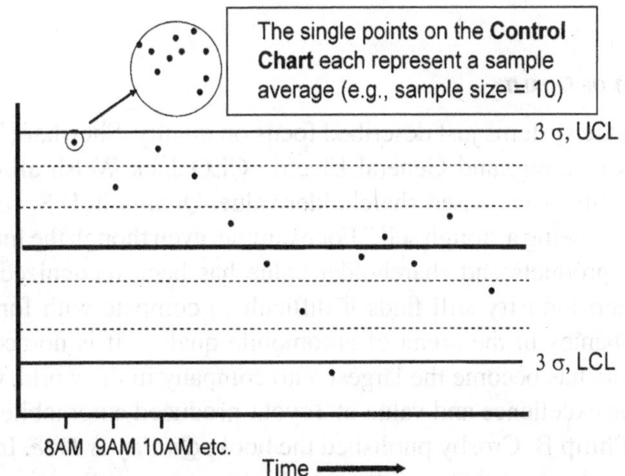

Fig. 9.6 Control chart.

process control methodologies that were able to reduce CTQ defects to 3.4 defects per million opportunities. Part of the underpinning of Six Sigma is statistical process control, but the technique emphasizes disciplined controls and mistake-proofing processes [8].

9.10.5 Toyota Production System

The Toyota Production System (TPS) [9] incorporates 14 principles that increase efficiency. At least one of those 14 principles confronts quality directly: "Principle 5, Build a culture of stopping to fix problems, to get quality right the first time." The TPS focuses on elimination of waste and categorizes quality defects as *waste that requires correction,* one of the seven kinds of "muda" (waste).

The lean techniques developed by Sakaichi Toyoda, Kiichiro Toyoda, and Taiichi Ohno of Toyota are often merged with Six Sigma to form *Lean Six Sigma*, an approach that has produced excellent results. Lean Six Sigma also has been adopted by all of the military services within the DoD; there are many documented examples of its benefit.

9.10.6 Malcolm Baldridge National Quality Award

The Baldridge National Quality Improvement Act of 1987 created the Malcolm Baldridge National Quality Award. This award competition was intended to stimulate quality and productivity within the United States to answer the challenge of foreign competition (competition that has exhibited more rapid productivity growth than has this country). This effort recognizes that poor quality costs companies as much as 20% of sales revenues and that improved quality improves productivity, reduces cost, and increases profitability.

9.10.7 Cost of Quality

Although the systems just described focus on quality, Shewhart, TQM guru W. Edwards Deming, and General Electric CEO Jack Welsh all emphasize corporate profits or increased shareholder value. Quality in U.S. corporations has a history of being a "tough sell." For example, even though the link between high-quality products and shareholder value has been recognized for many years, the auto industry still finds it difficult to compete with foreign automobile companies in the arena of automobile quality. It is no accident that Toyota Motors has become the largest auto company in the world. Customers recognize the excellence and value of Toyota-produced automobiles.

In 1979, Philip B. Crosby published the book, *Quality is Free*. In his book, Crosby made the case that quality mistakes were very expensive to correct and that it would cost less money to tighten quality controls to produce items right the first time. Crosby's point has been validated by the successes of SPC, TPS, and Six Sigma—all of which control processes within a narrower band than is required by the design tolerance. Controlling product variation costs less than rework, repair, or scrapping a defective item—counterintuitive as that sometimes seems.

The "hidden factory" is the notion that much of the company's endeavor that is not quality minded is directed inadvertently to creating waste and performing wasteful tasks. Examples of wasteful activities are the production of nonconforming products and the holding of excessive stock. The hidden factory is the extra useful, positive output that would theoretically be possible if the company's energy directed at creating waste were released and directed instead at making good-quality items. In 1977, the quality guru Armand Feigenbaum estimated the endeavor within the hidden factory might be 15 to 40% of total company effort. The notion of the hidden factory is bound up with the metric COPQ (cost of poor quality). The COPQ can be estimated by multiplying the number of defects per period of time by the average unit cost to fix a defect (labor and materials). Such a calculation, however, omits such costs as loss of goodwill and loss of competitiveness and other matters such as warranty costs and even legal damages (data available online at http://www.isixsigma.com/dictionary/Hidden-Factory—The-512.htm).

9.11 DoD POLICIES RELATED TO QUALITY

Current DoD guidance [10] states that contractors should have responsibility for the quality of their products. Program managers should allow contractors to define and use their own preferred quality-management systems, as long as those systems meet programs' required support capabilities. In general, contractors' quality-management systems should be capable of the following activities: monitor, measure, analyze, control, and improve processes; reduce product variation; measure/verify product conformity; establish mechanisms for field product performance feedback; and implement an effective root-cause analysis and corrective action system.

Federal Acquisition Regulation (*FAR*) Parts 9 and 46 [11] provide policy and guidance related to manufacturing and quality. *FAR* Part 9 (contractor qualifications) provides standards of responsibility for prospective contractors. A responsible prospective contractor is one that meets the standards in *FAR* 9.104, which states that, among other standards, a vendor must 1) have the necessary organization, experience, accounting and operational controls, and technical skills, or the ability to obtain them (including, as appropriate, such elements as production control procedures, property control systems, quality-assurance measures, and safety programs applicable to materials to be produced or services to be performed by the prospective contractor and subcontractors); and have the necessary production, construction, and technical equipment and facilities, or the ability to obtain them ([10], 9.104).

If the acquiring organization does not have sufficient information to determine prospective contractor responsibility, the organization can perform a preaward survey on a prospective vendor to determine if it has adequate technical, production, and quality-assurance capability for the proposed contract.

FAR 46 (quality assurance) provides policy and guidance on contract quality and quantity requirements to include inspection, acceptance, warranty, and other measures associated with quality requirements. It identifies four types of contract quality requirements:

1) In contracting for commercial items, the government relies on the contractors' existing quality-assurance systems as a substitute for government inspection and testing.

2) For supplies or services acquired at or below the simplified acquisition threshold, the government relies on the contractor to complete all necessary inspection and testing to ensure that the supplies or services conform to contract quality requirements.

3) Standard inspection requirements mandate the contractor provide and maintain an inspection system that is acceptable to the government. They also grant the government the right to make inspections and tests while work is in process and require the contractor to keep and make available to the government complete records of its inspection work.

4) Higher-level quality standards are appropriate in contracts for complex or critical items and when the technical requirements of the contract require either control of such things as work operations, in-process controls, and inspection, or attention to such factors as organization, planning, work instructions, documentation control, and advanced metrology ([10], 46.202).

9.12 CONCLUSION

Manufacturing and quality activities for major acquisition programs are complicated, detailed, and require the intricately integrated effort of thousands of government and contractor personnel. Production and quality must be carefully planned in advance throughout earlier preacquisition and acquisition phases before manufacturing processes can commence. As manufacturing gets underway, production personnel must expend significant effort to ensure that manufacturing processes lead to high-quality products that satisfy the needs of military users. At the completion of the production process, warfighting systems of known quality are deployed to the Armed Forces. America's warfighting systems must perform as required; both the lives our service personnel and the defense of our nation depend on them.

REFERENCES

[1] Government Accountability Office, "Best Practices: Capturing Design and Manufacturing Knowledge Early Improves Acquisition Outcomes," USGAO-02-701, Washington, D.C., July 2002.

[2] Dept. of Defense, *"Technology Risk Assessment (TRA) Deskbook"*, Washington, D.C., May 2005.

[3] Dept. of Defense, "Transition from Development to Production," DoD 4245.7-M, Washington, D.C., Feb. 1985.

[4] Boothroyd, G., Dewhurst, P., and Knight, W., *Product Design for Manufacture & Assembly*, Marcel Dekker, New York, 1994.

[5] Womack, J. P., and Jones, D. T., *Lean Thinking: Banish Waste and Create Wealth in Your Corporation*, Free Press, New York, 2003.

[6] Goetsch, D. L., and Davis, S. B., *Quality Management: Introduction to Total Quality Management for Production, Processing, and Services*, 3rd ed., Prentice-Hall, Upper Saddle River, NJ, 2000.

[7] Harry, M., and Schroeder, R., *Six Sigma: The Breakthrough Management Strategy Revolutionizing the World's Top Corporations*, Currency, New York, 2000, p. 9.

[8] Harry, M., and Schroeder, R., *Six Sigma: The Breakthrough Management Strategy Revolutionizing the World's Top Corporations*, Currency, New York, 2000, p. 35–41.

[9] Liker, J. K., *The Toyota Way*, McGraw-Hill, New York, 2003.

[10] Dept. of Defense, *Defense Acquisition Guidebook*, Defense Acquisition University Website, https://akss.dau.mil/dag/[retrieved 22 July 2007].

[11] General Services Administration, *Federal Acquisition Regulation*, Washington, D.C., March 2005.

STUDY QUESTIONS

9.1 What is the relationship of manufacturing and quality to the systems-engineering process?

9.2 What is producibility?

9.3 What producibility development activities occur in concept development? In preliminary design and detailed design?

9.4 What is the meaning of knowledge points 1, 2, and 3? Can you relate the meaning of each to manufacturing?

9.5 What are three systems-engineering reviews that address production and quality readiness? Explain how each influences production.

9.6 What are three risk drivers that cause concern in determining if a system is ready for production? [Hint: A good way to describe risks is in the form of a question. For example, "Can we meet our production schedule?"] What are some metrics that can be used to track those same production risk drivers?

9.7 What are the five process steps that describe *lean*?

9.8 Which is the better quality approach: inspecting product at the end of the production process or carefully controlling those manufacturing processes that are critical to quality? What is your rationale?

9.9 What is the most prevalent industry standard quality system?

Chapter 10

Contract Management

Rene G. Rendon*
Naval Postgraduate School, Monterey, California

Learning Objectives

- Explain how contract management is an essential part of the management of defense acquisition projects
- Discuss the sources of contracting authority for government contracting officers
- Identify some of the contracting regulations used in defense contracting
- List some of the statutory requirements and some common government socioeconomic programs that are supported by the defense contracting process
- Explain the roles and responsibilities of the contracting officer, program manager, and contractor personnel, as well as the relationship between the contracting officer and other members of the project team
- Understand the six phases of the contract-management process, from both the government and contractor perspectives
- Describe the contracting activities involved in each of the six phases of the contract-management process, from both the government and contractor perspectives
- Understand some of the statutory and regulatory requirements affecting the six phases of the contract-management process
- Describe some of the emerging trends and current challenges and opportunities in defense contracting

---

*Senior Lecturer, Graduate School of Business and Public Policy.

10.1 INTRODUCTION

It has often been said that "accounting is the language of business" ([1], p. 1). If this is true, then it could also be argued that "contracting is the language of defense acquisition." Just as business managers and organizations use accounting processes, concepts, and terms to manage and describe everyday business activities, defense acquisition managers, both on the buying side as well as on the selling side, use contracting processes, concepts, and terms, to manage and describe the activities and events that occur in defense acquisition programs. The purpose of this chapter is provide an overview of the contract management processes, concepts, and terms that are used by defense acquisition managers, defense organizations, and defense contractors in managing defense acquisition programs. The focus will be on the contract management processes encompassing both pre-award and post-award processes, as performed by both government and contractor organizations. One of the goals of this chapter is to present contract management as an essential part of the management of defense acquisition projects, both for the Defense Department as well as for the defense contractor organizations.

First, a discussion of basic government contract concepts will be presented with a review of sources of contracting authority, contracting regulations, statutory requirements, and some common government socioeconomic programs. The roles and responsibilities of the contracting officer, program manager, and contractor personnel will then be discussed, as well as the relationship between the contracting officer and other members of the project team. The contract-management process will then be presented, focusing on the common phases and activities that occur during the contract-management process, from both the government and contractor perspective. Finally, emerging trends in defense contracting will be presented, along with highlights of current challenges and opportunities. References will be made both to statutory requirements and federal regulations pertaining to various aspects of the contract-management process, as well as to current contracting trends and issues occurring in today's defense acquisition environment.

10.2 GOVERNMENT CONTRACTING BASICS

The Federal Acquisition System includes the processes, policies, and regulations used by the federal government for obtaining the supplies and services needed to perform its mission. As stated in the *Federal Acquisition Regulation*, "the vision for the Federal Acquisition System is to deliver on a timely basis, the best value product or service to the customer, while maintaining the public's trust and fulfilling public policy objectives" [2]. It is this vision of the Federal Acquisition System that makes it unique from commercial contracting processes. The use of the contracting process by the federal government as a vehicle for achieving public policy objectives definitely differentiates the federal contracting process from contracting processes used by commercial and nongovernmental entities.

10.2.1 SOURCES OF AUTHORITY

The Constitution of the United States provides the legal authority for the government to enter into contracts for supplies and services. This authority is derived from Article 1, Section 8. Clause 12 of this section specifically states that Congress shall have the power to raise and support armies, and Clause 13 specifically states that Congress shall provide and maintain a navy. (In 1948, the federal government determined that the Air Force could be considered an army within the meaning of Article 1, Section 8, Clause 12.)

The basic provisions of the Constitution are implemented through the statutes and laws that have been formulated by the government. These laws are codified in the United States Code (U.S.C.), and the laws specifically related to the Armed Forces are found in Title 10 of the U.S. Code. For example, the Competition in Contracting Act of 1984 is found in 10 U.S.C. 2304, as well as 41 U.S.C. 253.

In addition to the Constitution, and the codification of its implementing statutes and laws in the United States Code, there are also numerous executive orders, regulations, and directives that govern the federal contracting process. The primary contracting regulation is the *Federal Acquisition Regulation*, which will be discussed next.

10.2.2 FEDERAL ACQUISITION REGULATION

The *Federal Acquisition Regulation (FAR)* provides uniform policies and procedures for acquisition by all executive agencies. In addition, agencies implement or supplement the *FAR* with their own agency regulations. For example, the *Department of Defense FAR Supplement (DFARS)* implements or augments the *Federal Acquisition Regulation* for the U.S. Department of Defense (DoD). Furthermore, within the Department of Defense, the military services also maintain regulations that implement or supplement the *Federal Acquisition Regulation*. The *FAR* is available online at www.arnet.gov/far.

10.2.3 STATUTORY REQUIREMENTS

As just discussed, the Competition in Contracting Act of 1984 is one of the many federal statutes that affect government contracting. The Competition in Contracting Act (CICA) is a public law enacted to increase the number of government procurements conducted using the procedures of full and open competition. Unless the government can justify an exception to the CICA requirements, the procurement must provide for full and open competition in the solicitation and award of the contract.

Another example of a statutory requirement for government contracting is the Truth in Negotiations Act (TINA) found at 10 U.S.C. 2306. The Truth in

Negotiations Act provides for full and fair disclosure by contractors in the conduct of negotiations with the government. The TINA requires that contractors submit certified cost and pricing data for specific negotiated procurements above a certain dollar threshold.

The legislation formulated by Congress to govern the federal contracting process is one of the characteristics that makes government contracting unique. Another area that is unique to government contracting is the support of socioeconomic programs. The following are just a few of the various socioeconomic programs supported by government contracting regulations:

Buy American Act: The Buy American Act sets a preference for domestic materials over foreign materials.

Javits–Wagner O'Day Act (*JWOD*): This act provides employment opportunities for people who are blind or have other severe disabilities in the manufacture and delivery of products and services to the federal government.

Federal Prison Industries: This program requires the purchase of specific supplies from Federal Prison Industries, Inc.

Small Business Act: This act requires awarding a fair portion of government contracts to small businesses, as defined by the *Federal Acquisition Regulation.*

10.3 Roles and Responsibilities in Defense Contract Management

The Statement of Guiding Principles in the *Federal Acquisition Regulation* states that the acquisition team consists of "all participants in Government acquisition including not only representatives of the technical, supply, and procurement communities but also the customers they serve, and the contractors who provide the products and services" [2]. The acquisition team represents the application of the project team concept (discussed in Chapter 1) to the management of defense acquisition projects.

This project team is a cross-functional team that represents all participants in the defense contract-management process. These project team members include the program or project manager (PM) and the supporting functional representatives who are part of the acquisition process. These functional representatives typically include the contracting officer, financial manager, technical representative, and other functional representatives who are involved in the specific acquisition. In addition, the project team also includes the organization's customers and the contractors who provide professional and administrative services to the organization, as well as the prime contractor providing the supplies or services required of the project.

The PM is the individual who is solely responsible for leading the project team in support of the projects cost, schedule, and performance objectives.

With the help of the various functional members of the project team, the PM acts as a facilitator, coordinator, and integrator of the various activities being performed in support of the project objectives. An analogy often used to portray the PM and the project team is that of an orchestra conductor. The PM acts as an orchestra conductor ensuring that each member of the orchestra (project team) is in tune with the musical score (project plan) and is successfully performing in their specific area of expertise (functional area) ([4], p. 1). In many defense acquisition programs, there are typically two PMs: the PM leading the government team and the PM leading the contractor team. The government project team and the contractor project team typically have organizations that mirror each other.

In defense contract management, one of the critical members of the project team is the government contracting officer. The contracting officer is responsible for providing the contracting support required for meeting the project's cost, schedule, and performance objectives. Additionally, the government contracting officer is responsible, not only for providing contracting support to the project team, but also for ensuring that the project's contracting activities are in compliance with the statutory requirements and agency regulations. This typically places the government contracting officer in a pivotal position: challenged with supporting project activities as well as ensuring compliance with statutory requirements and agency regulations. Box 10.1 illustrates this aspect of the contract manager's position. More important, the government contracting officer is the only individual who has the authority to enter into, administer, or terminate contracts and make contracting-related determinations and findings ([2], Part 1.602). Just as in the case of the government PM, the government contracting officer also has a counterpart on the contractor team—the contractor's contract manager. The formal (as well as informal) contracting documentation and correspondence flows from the government contracting officer to the contractor's contract manager.

BOX 10.1 THE CONTRACTS MANAGER

Contracts management is a team activity in which the contracts manager is and should be an unpopular player ([4], p. 4).

Another critical member of the government's project team is the government finance manager. The finance manager ensures that the project team is complying with the appropriate financial management regulations (FMR) in terms of ensuring adequate and appropriate funding is available for the project's contracts. This includes monitoring the project team's compliance with the statutory requirements concerning purpose, time, and amount restrictions.

10.4 CONTRACT-MANAGEMENT PROCESS

Typically, contract management is discussed from the perspective of the buyer, with a focus on the procurement (buying) side of contracting. However, the mirror side of procurement (that is, the selling side), represented by defense contractors, is equally as critical to the contract-management process. As stated under roles and responsibilities, contract managers exist in both government (buying) as well as contractor (selling) organizations; both are concerned with managing the contractual aspects of defense acquisition programs.

Contract management can be defined as "the art and science of managing a contractual agreement throughout the contracting process" ([5], p. 390). As stated earlier, the contract-management process encompasses both pre-award and post-award processes as performed by both government and contractor organizations. Thus, contract-management activities are performed by both the buyer and seller organizations. Because this book is about management of the defense acquisition projects (which occur both in buyer and seller organizations), this chapter on contract management will focus on the contract-management processes performed by both buyer and seller organizations. The contract-management process can be analyzed using a six-phase model, as illustrated in Fig. 10.1.

These six phases for the buying organization consist of procurement planning, solicitation planning, solicitation, source selection, contract administration, and contract closeout. For the seller's organization, the six contract-management phases consist of presales activity, bid/no-bid decision making, bid or proposal preparation, contract negotiation and formation, contract administration, and contract closeout [5]. As you can see, both the buyer and

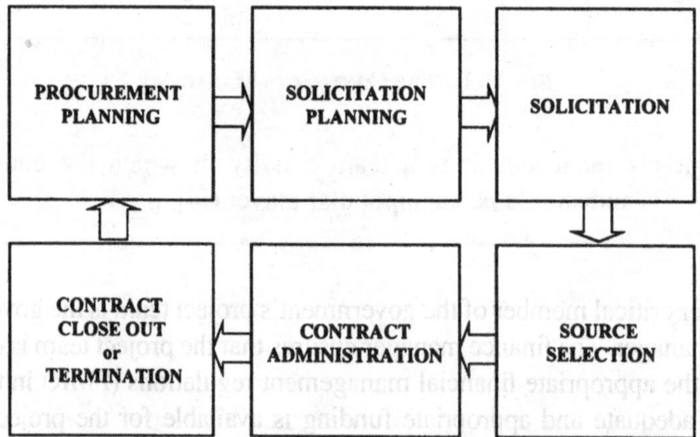

Fig. 10.1 Contract-management process (buyer's perspective).

seller perform similar contract administration and contract closeout functions during the management of the contractual relationship. Indeed, the contract-management processes of contract administration and contract closeout are identical for both the buying and selling organizations. The seller's contract management activities will be discussed further later in the chapter.

As reflected in Fig. 10.2, the contract-management process is typically used as the vehicle for progressing through the defense acquisition life cycle. Using this textbook's framework for representing the various phases and milestones of a defense weapons system acquisition program, we can see how the contract-management process is used to plan, award, administer, and close out contracts for the work required in each of the phases of the project life cycle—need development, concept development, preliminary design, detailed design, and production, deployment, operations, support, and disposal.

Each of these contract management phases provides critical planning, execution, and control of the overall contracting process and is integral to the success of the resultant contract and contractor performance. This chapter will provide a brief overview of the contract-management process, from the buyer and seller perspectives, and identify key areas for consideration for each contracting phase.

10.4.1 PROCUREMENT PLANNING

Procurement planning involves "the process of identifying which business needs can be best met by procuring products or services outside the

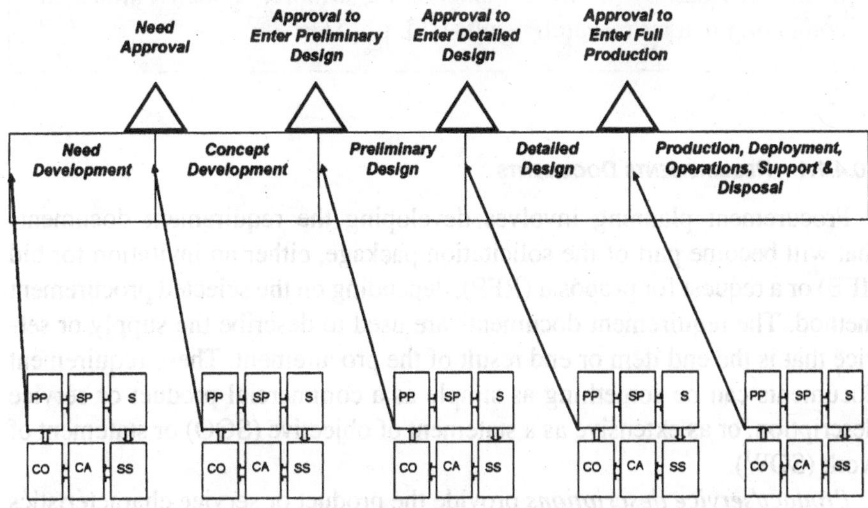

Fig. 10.2 Contracts and the project life cycle.

organization. This process involves determining whether to procure, how to procure, what to procure, how much to procure, and when to procure" ([5], p. 81). This phase of the contracting process includes 1) conducting outsource analysis; 2) determining and defining the requirement (the supply or service to procure); 3) conducting market research and/or a presolicitation conference; 4) developing preliminary requirements documents such as work breakdown structures (WBS), statements of work (SOW), performance work statement (PWS), or other descriptions of the supply or service to be procured; 5) developing preliminary budgets and cost estimates; 6) preliminary consideration of contract type and consider any special contract terms and conditions; and 7) conducting risk analysis. These procurement planning activities will determine the contracting strategy to be used for the defense acquisition program. Thus, the results of procurement planning and its related activities must be sufficiently documented and may require review and approval at higher levels of the agency. Box 10.2 illustrates the importance of sufficient documentation of procurement planning activities, specifically the market research activities.

BOX 10.2 SUFFICIENT DOCUMENTATION

Marine Corps Systems Command (MCSC) contracting officials stated that the sole-source award to a contractor for armored vehicles was based, in part, on the results of adequate market research. However, MCSC officials were unable to provide sufficient documentation to support their market research results. In addition, market research documents provided by MCSC officials on the armored vehicles indicated conflicting market research results ([6], p. 15).

10.4.1.1 REQUIREMENTS DOCUMENTS

Procurement planning involves developing the requirement documents that will become part of the solicitation package, either an invitation for bid (IFB) or a request for proposal (RFP), depending on the selected procurement method. The requirement documents are used to describe the supply or service that is the end item or end result of the procurement. These requirement documents can be something as simple as a commercial product or service description, or as extensive as a statement of objective (SOO) or statement of work (SOW).

Product/service descriptions provide the product or service characteristics required by the procurement.

Specifications describes the technical requirements for items, materials and services, including the procedures by which it will be determined that the requirements have been met [7]. Specifications can be design specifications or performance specifications.

Statement of objectives provides a broad description of the government's required performance objectives [7].

Statement of work provides all nonspecification requirements for contractors efforts either directly or with use of specific cited documents [7].

It should be noted that *FAR* 11.01 provides an order of precedence for selecting requirements documents to meet agency needs as follows: 1) documents mandated for use by law; 2) performance-oriented documents (e.g., a PWS or SOO); 3) detailed design-oriented documents; and 4) standards, specifications, and related publications issued by the government outside the defense or federal series for the nonrepetitive acquisition of items [2].

10.4.2 SOLICITATION PLANNING

Solicitation planning involves the "process of preparing the documents needed to support the solicitation. This process involves documenting program requirements and identifying potential sources" ([5], 405). This process includes 1) determining procurement method (sealed bids, negotiated proposals, e-procurement methods, procurement cards, etc.); 2) determining contract type (fixed price vs cost type); 3) developing the solicitation document (IFB, RFQ, or RFP); 4) determining proposal evaluation criteria and contract-award strategy; 5) structuring contract terms and conditions; and 6) finalizing solicitation WBS, SOW, or product or service descriptions.

PROCUREMENT METHODS. In defense contracting, there are two major procurement methods; sealed bidding and negotiated procurement. Sealed bidding is a method of contracting that employs competitive bids, public opening of bids, and awards. As described in *FAR* 14, sealed bidding is used when there is adequate time for the solicitation, submission, and evaluation of sealed bids; the contract award will be made on the basis of price and other price-related factors. It is not necessary to conduct discussions with the responding offerors about their bids, and there is a reasonable expectation of receiving more than one sealed bid ([2], Part 6.401).

Negotiated procurement, using procedures in *FAR* 15, is used when sealed bids are not appropriate—that is, when the contract award might not be made on the basis of price and other price-related factors, and it might be necessary to conduct discussions with the responding offerors about their proposals.

The other procurement methods, e-procurement and procurement cards, are used for smaller-dollar and routine purchases by the Department of Defense and other federal agencies.

Competition Requirements. Contracting by negotiation can be conducted in either a sole-source environment or in a competitive environment. The CICA requires contracting officers to promote and provide for full and open competition in soliciting offers and awarding government contracts. CICA allows for other than full and open competition given the following circumstances ([2], Part 6.302): 1) only one responsible source, and no other supplies or services will satisfy agency requirements; 2) unusual and compelling urgency; 3) industrial mobilization, engineering, developmental, or research capability, or expert services; 4) international agreement; 5) authorized or required by statute; 6) national security; or 7) public interest. Box 10.3 illustrates the competition strategy for the Air Force KC-X Program.

> **Box 10.3 Full and Open Competition**
>
> "To ensure full and open competition and to eliminate impediments to competition, the acquisition strategy for the KC-X Program provided for the Air Force to competitively contract for as many as 80 commercial derivative aircraft of the intended procurement of 179 KC-X tanker aircraft. Consistent with *FAR*, Subpart 6.101, the KC-X Program Office promoted full and open competition in soliciting offers in its request for proposal for as many as 80 commercial derivative aircraft, as described in the KC-X acquisition strategy" ([8], p. 5).

Types of Contracts. The *Federal Acquisition Regulation* categorizes the major contract types as fixed price, cost reimbursement, incentive contracts, and indefinite delivery contracts.

Fixed-price contracts can provide for either a firm price or an adjustable price. "Fixed-price contracts providing for an adjustable price may include a ceiling price, target price (including target cost), or both" ([2], Part 16.201). The major types of fixed-price contracts include firm fixed price, fixed-price contracts with economic price adjustment, fixed-price incentive contracts, fixed-price contracts with prospective price redetermination, fixed-ceiling-price contracts with retroactive price redetermination, and firm-fixed-price, level-of-effort term contracts ([2], Part 16.2).

Cost-reimbursement contracts provide for payment of allowable incurred costs as prescribed in the contract. These contracts establish "an estimate of total cost for the purpose of obligating funds and establishing a ceiling that the contractor may not exceed (except at its own risk) without the approval of the contracting officer" ([2], Part 16.301). The major types of cost-reimbursement contracts include cost contracts, cost sharing, cost-plus-incentive-fee, cost-plus-award-fee, cost-plus-fixed-fee ([2], Part 16.3).

BOX 10.4 AWARD FEES

A recent GAO reported stated that DoD acquisition programs "regularly paid contractors a significant portion of the available fee for what award-fee plans describe as 'acceptable, average, expected, good, or satisfactory' performance, when federal acquisition regulations and military service guidance state that the purpose of these fees is to motivate excellent performance. These practices reduce the effectiveness of award fees as motivators of performance and compromise the integrity of the fee process" ([9], p. 3).

Incentive contracts are appropriate when a firm fixed-price contract is not appropriate and required supplies or services can be acquired both at lower cost and (in certain instances with improved delivery or technical performance) by relating the amount of profit or fee payable under the contract to the contractor's performance ([2], Part 16.401). Incentive contracts can be crafted based on objective predetermined, formula-type incentives in the areas of cost, performance, and/or delivery. Award fee contracts are also considered a type of incentive contract. Award fees paid to the contractor are based on the government's judgmental evaluation of the contractor's performance. Boxes 10.4 and 10.5 discuss recent GAO reports on DoD's use of award fee contracts.

The *FAR* identifies three types of *indefinite-delivery contracts*: definite-quantity contracts, requirements contracts, and indefinite-quantity contracts. The appropriate type of indefinite-delivery contract can be used to acquire supplies and/or services when the exact times and/or exact quantities of future deliveries are not known at the time of contract award ([2], Part 16.501-2). A definite-quantity contract provides for delivery of a definite quantity of specific supplies or services for a fixed period, with deliveries or performance to be scheduled at designated locations upon order ([2], Part 16.502). A requirements contract provides for filling all actual purchase requirements of designated government activities for supplies or services during a specified contract period, with deliveries or performance to be scheduled by placing orders with the contractor ([2], Part 16.503). An indefinite-quantity contract provides for an indefinite quantity, within stated limits, of supplies or services during a fixed period. The government places orders for individual requirements. Quantity limits can be stated as number of units or as dollar values ([2], Part 16.504).

Each of the various types of contracts (specifically cost-reimbursement vs fixed-price contracts) reflects the sharing and allocation of risk between the buyer and the seller. As illustrated in Fig. 10.3, fixed-priced contracts are

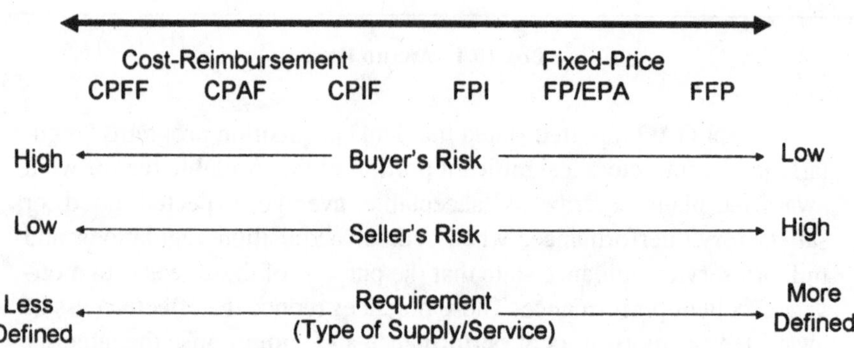

Fig. 10.3 Relationship between contract type, requirement, and risk.

appropriate for well-defined requirements in situations with a low risk of performance. Using these types of contracts, the contractor holds the major burden of risk. On the other hand, under cost-reimbursement contracts, which are appropriate for developmental requirements, the risk of performance is high. In these types of contracts, the government shares the major burden of risk.

UNIFORM CONTRACT FORMAT. The *Federal Acquisition Regulation* provides a uniform contract format to facilitate the "preparation of the solicitation and contract, as well as reference to, and use of, those documents by offerors, contractors, and contract administrators" ([2], Part 15.204). The uniform contract format is shown in Fig. 10.4. The uniform contract format for solicitations and contracts consists of four parts containing a total of 13 sections ([2], Part 15.204). The following discussion on the uniform contract format is summarized from *FAR* 15.204.

Box 10.5 COST-PLUS-AWARD-FEE CONTRACTS

A recent GAO study determined that on cost-plus-award-fee contracts, the award fee is often the only source of potential fee for the contractor. According to defense acquisition regulations, these contracts can include a base fee—a fixed fee for performance paid to the contractor—of anywhere from 0 to 3% of the value of the contract; however, based on GAO sample results, about 60% of the cost-plus-award fee contracts in the study population included zero base fee. However, the two F/A-22 development contracts in the GAO sample included a 4% base fee. The program office received a deviation from the *Defense Federal Acquisition Regulation Supplement*, which now allows for a maximum of 3% base fee ([12], p. 7).

Section Title

Part I – The Schedule

A: Solicitation/contract form.
B: Supplies or services and prices/costs.
C: Description/specifications/statement of work.
D: Packaging and marking.
E: Inspection and acceptance.
F: Deliveries or performance.
G: Contract administration data.
H: Special contract requirements.

Part II – Contract Clauses

I: Contract clauses.

Part III – List of Documents, Exhibits, and Other Attachments

J: List of attachments.

Part IV – Representations and Instructions.

K: Representations, certifications, and other statements of offerors or respondents.
L: Instructions, conditions, and notices to offerors or respondents.
M: Evaluation factors for award.

Fig. 10.4 Uniform contract format (*FAR* 15.204).

Part I is the schedule and consists of Secs. A through H.

Section A is the solicitation/contract form and consists of the cover page of the solicitation/contract document. The cover page provides basic information such as name, address, and location of issuing activity, solicitation number, date of issuance, closing date, brief description of the item or service, and offer expiration date. Section A also indicates if the contract is a rated order under the Defense Priorities and Allocation System (DPAS).

Section B is the supplies or services and prices/costs. This section contains a brief description of the supplies or services, quantities, unit prices, and total prices. In addition, this section also identifies the contract line item numbers (CLINs) as well as the type of contract for each CLIN. A typical contract might consist of a number of different types of contract relationships. A contract might include a number of different CLINs, with each CLIN having a different type of contract, such as firm-fixed-price (FFP), fixed-price incentive (FPI), cost-plus-fixed-fee (CPFF), cost-plus-award-fee (CPAF), and so forth.

Section C is the description/specification/statement of work. This section includes additional descriptions or specifications for supplies or services identified in Sec. B. Although the statement of work is referenced in Sec. C, the SOW is typically physically located as an attachment in Sec. J.

Section D is packaging and marking. This section contains packaging, packing, preservation, and marking requirements, if any, for the supplies or services to be delivered.

Section E is inspection and acceptance. This section contains inspection, acceptance, quality assurance, and reliability requirements for the supplies or services to be delivered.

Section F is deliveries or performance. This section specifies the requirements for time, place, and method of delivery or performance for the supplies or services.

Section G is contract administration data. This section provides any required accounting and appropriation data and any required contract administration information or instructions, other than those listed in Sec. A.

Section H is special contract requirements (SCRs). This section includes special contract requirement clauses, also known as Sec. H clauses. Section H clauses are typically unique and specific clauses developed by the local agency and tailored to that particular acquisition. Section H clauses are developed for any special contract requirement that is not included in the general Sec. I contract clauses or in other sections of the uniform contract format.

Part II is contract clauses and consists of Sec. I.

Section I is composed of contract clauses and includes the clauses required by law or by the *Federal Acquisition Regulation.* It also contains any additional clauses expected to be included in any resulting contract, if these clauses are not required in any other section of the uniform contract format. *FAR* Part 52 provides instructions for using the provisions and clauses in solicitations and contracts and also sets forth solicitation provisions and contract clauses prescribed by the *FAR*. In addition, *FAR* 52.3 provides a provisions and clauses matrix, which lists every provision or clause of the *FAR*, the prescription of the provision or clause, and whether the provision or clause is required, required when applicable, or optional for each principle type of contract.

Part III is a list of documents, exhibits, and other attachments and consists of Sec. J.

Section J is a list of attachments and contains the title, date, and number of pages for each attached document, exhibit, and other attachment. Examples of documents, exhibits, and other attachments include the contract WBS, SOO or SOW, government-furnished property (GFP) listing, specifications listing, and the contract data requirements list (CDRL).

Part IV is representations and instructions and contains Secs. K, L, and M. Section K is representations, certifications, and other statements of offerors. This section includes solicitation provisions that require representations, certifications, or the submission of other information by offers. Examples of these types of representations and certifications relate to certificate of independent price determination, taxpayer identification, women-owned business, and affirmative action compliance. Section L is instructions, conditions, and notices to offerors or respondents. This section contains solicitation provisions and other information and instructions to guide offerors or respondents in preparing proposals or responses to requests for information. This section provides instructions on formatting and organization requirements of contractor proposals.

Section M is evaluation factors for award. This section provides all significant factors (and any significant subfactors that will be considered in awarding the contract) and their relative importance. It is in this section that the solicitation will state if contract award will be made to the lowest price technically acceptable proposal or to other than the lowest priced technically acceptable proposal, and whether the government intends to conduct discussions during source selection.

Once the contract is awarded (signed by the contractor first and then the government), Part IV of the solicitation (representations and instructions containing Secs. K, L, and M) is physically removed from the contract and retained in the contract files. The award of the contract includes incorporating Sec. K by reference in the resultant contract.

10.4.3 SOLICITATION

Solicitation is the process of obtaining information (bids and proposals) from the prospective sellers on how project needs can be met ([5], p. 90). This process includes 1) conducting advertising of the procurement opportunity, or providing notice to interested offerors; 2) conducting pre-proposal conferences, if required; and 3) developing and maintaining a qualified bidder's list. The importance of advertising government contracting opportunities during the solicitation phase is to increase competition, broaden industry participation in meeting government requirements, and assist the various types of small business concerns in obtaining contracts and subcontracts (*FAR* 5.002). As you recall from this chapter's section on statutory requirements, the Competition in Contracting Act of 1984 requires contracting officers to provide for full and open competition in the solicitation and award of contracts. Exceptions to competition are allowed under the CICA and are listed in *FAR* 6.302.

Federal government contracting opportunities are publicized through the Government Point of Entry. When the *FAR* requires that a contracting opportunity notice be published, the contracting officer must transmit the notice to the Government Point of Entry. The Government Point of Entry (GPE) is the single point where government business opportunities greater than $25,000, including synopsis of proposed contract actions, solicitations, and associated information, can be accessed electronically by the public. The GPE is located at http://www.fedbizopps.gov ([2], Part 2.101).

10.4.4 Source Selection

Source selection is the process of "receiving bids or proposals and applying the proposal evaluation criteria to select a supplier" ([5], p. 137). The source-selection process includes the evaluation of offers and proposals and contract negotiations between the buyer and the seller in attempting to come to agreement on all aspects of the contract, including cost, schedule, performance, terms, and conditions, and anything else related to the contracted effort. This process includes 1) applying evaluation criteria to management, cost, and technical bids or proposals; 2) negotiating with suppliers; and 3) executing the contract award strategy.

The complexity of the source-selection process will depend on the procurement method selected. The source-selection process for a sealed-bidding procurement will be formal, structured, and very mechanical, with the award being made to that responsible bidder whose bid, conforming to the invitation, will be most advantageous to the government, considering only price and the price-related factors ([2], Part 14.408-1). In this type of source selection, there are no discussions or contract negotiations, and the types of contract used is typically firm-fixed-price, except that fixed-price contracts with economic price-adjustment clauses can be used if authorized.

Negotiated procurements, on the other hand, entail more of an extensive and complex source-selection approach, especially if a tradeoff process is used. In these types of procurements, the source-selection process can include oral presentations, exchanges with offerors (to include clarifications and communications), as well as a requirement for submission and certification of cost or pricing data.

Best-Value Continuum. As stated in the *FAR*, "the vision for the Federal Acquisition System is to deliver on a timely basis the best value product or service to the customer, while maintaining the public's trust and fulfilling public policy objectives" ([2], Part 1.102). The *FAR* describes "best value" as a continuum in which the relative importance of cost or price can vary for each specific procurement situation. In some contract source selections, in which the requirement is clearly definable and the risk of unsuccessful contract performance is minimal, cost or price can play a dominant role. In other source selections, in which the requirement is less definitive and more development work is required (resulting in greater performance risk), more technical or past performance considerations can play a dominant role ([2], Part 15.101).

Lowest Price Technically Acceptable and Tradeoff Process. The government uses the lowest price technically acceptable source-selection process when best value is expected to result from selection of the technically acceptable proposal with the lowest evaluated price. In a lowest price technically acceptable source selection, the solicitation provides the evaluation factors and significant subfactors that establish the requirements of acceptability.

In addition, tradeoffs are not permitted, and proposals are evaluated for acceptability but are not ranked using the non-cost/price factors ([2], Part 15.101-2). The contract is awarded to the proposal considered technically acceptable and also determined to be the lowest price.

The government uses the tradeoff process when it is believed that best value can be obtained from an "award to other than the lowest priced offeror or other than the highest technically rated offeror" ([2], Part 15.101). The tradeoff process allows the government the flexibility to award to an offeror anywhere on the best-value continuum between the *lowest priced* technically acceptable offeror and the *highest technically* rated offeror. "This process permits tradeoffs among cost or price and non-cost factors and allows the Government to accept other than the lowest priced proposal" ([2], Part 15.101). In this type of source selection, the solicitation shall clearly state all evaluation factors and significant subfactors that will affect contract award and their relative importance. In addition, the solicitation shall state whether all evaluation factors other than cost or price, when combined, are significantly more important than, approximately equal to, or significantly less important than cost or price ([2], Part 15.101).

SOURCE-SELECTION ORGANIZATION. For the more complex source selections, a formal source-selection organization is required to manage the source-selection process. The source-selection organization typically includes the source-selection authority, source-selection advisory council, source-selection evaluation team, and the contracting officer.

The source-selection authority is responsible for establishing the source-selection organization, approving the source-selection strategy plan, overseeing the contractor proposal evaluation process, and ultimately selecting the source or sources whose proposal is the best value for the government. Depending on the complexity and dollar value of the source selection, the contracting officer might, in fact, be the source-selection authority.

The source-selection advisory council includes seasoned and experienced cross-functional representatives that act as advisors and provide advice and recommendations to the source-selection authority.

The source-selection organization also includes a source-selection evaluation team. This team is a cross-functional team that includes appropriate representation from contracting, legal, logistics, technical, and other fields of expertise to ensure a comprehensive evaluation of contractor proposals. Because the source-selection evaluation team is responsible for the evaluation of contract proposals, it would also seem logical that the source-selection evaluation team would be instrumental in the development of the solicitation document that requests such proposals. Thus, the source-selection evaluation team should be formed during the procurement planning process to ensure consistency between the development of the solicitation document, the evaluation of contract proposals, and the award of the contract.

The contracting officer is responsible for serving as a focal point for inquiries from industry after release of the solicitation. Additionally, the contracting officer controls exchanges with offerors during the evaluation of proposals. And, ultimately, the contracting officer awards the contract.

10.4.5 Contract Administration

Once the contract is awarded, the contract-administration phase of the contract management process begins. Contract administration is the "process of ensuring that each party's performance meets the contractual requirements" ([5], p. 162). The activities involved in contract administration will depend on the contract statement of work, contract type, and contract performance period. The contract-administration process typically includes 1) conducting a preperformance conference, 2) monitoring the contractor's work results, 3) measuring contractor's performance, and 4) managing the contract change-control process.

Contract administration reflects the monitoring and controlling aspect of the project-management processes discussed in Chapter 1 of this text. The monitoring and controlling project-management processes are focused on ensuring that project objectives are met. To do this, the PM must perform such activities as measuring progress against plan-holding-status meetings and correcting the divergences from schedule or budget. During the contract-administration phase, the buyer and the seller are both focused on ensuring that each party's performance meets contractual requirements. In major defense acquisition projects, the contract-administration phase is critical to effective project management. It is this phase of the contract-management process in which the contractor is performing the statement-of-work (SOW) requirements, and the completed work is then measured and evaluated by the buying organization.

FAR Part 42, Contract Administration, provides policies and procedures for performing contract-administration functions on government contracts. This includes assigning contract-administration functions, managing contract correspondence, conducting preperformance conferences, and contractor production surveillance.

Pre-performance Conference. Just as in the kick-off of any project, the contract-administration phase should also start with a kick-off meeting. This kick-off meeting is the preperformance conference (sometimes called the postaward conference). This is typically the first time that the buying organization and the selected contractor meet after the announcement of contract award. The purpose of this conference is to ensure a clear and mutual understanding of all contract requirements and to identify and resolve any potential problems. It is at this conference that key personnel are identified, and roles and responsibilities are confirmed for both government and contractor personnel.

MONITORING THE CONTRACTOR'S PERFORMANCE. During the contract performance period, the contractor is hard at work performing the SOW requirements and other requirements of the contract. A significant aspect of the contract-administration phase consists of monitoring the contractor's performance. Depending on the type of contract (whether developmental, production, or services contract), the government typically uses quality-assurance evaluators (QAE), quality-assurance representatives (QAR), or contracting officer technical representatives (COTR) to perform the technical aspects of monitoring the contractor's performance. These technical representatives act as the contracting officer's eyes and ears in terms of ensuring the contractor meets the technical requirements of the contract. It is these technical representatives who will determine if the contractor is deficient in performing the contractual requirements, if the contractor is meeting the required standards, or if the contractor is exceeding the contractual requirements. It is also these technical representatives who can determine that the technical requirements documents, such as the SOW or specifications, need to be revised or corrected. Typically, any revisions to the technical requirements document will result in a formal modification to that specific section of the contract with any associated changes to costs or schedule requirements.

In major defense acquisition contracts, a designated contracting officer might be assigned specifically for administering the contract. These contracting officers are referred to as administrative contracting officers (ACOs) and have specifically delegated contract-administration functions as identified in the *Federal Acquisition Regulation* Part 42 [2].

DEFENSE CONTRACT MANAGEMENT AGENCY. In addition, major defense contracts might require the support from the Defense Contract Management Agency (DCMA). The DCMA is a Department of Defense combat-support agency ensuring the integrity of the contractual process and providing a broad range of acquisition management services for America's warriors. The DCMA provides customer-focused acquisition support and contract-management services to ensure warfighter readiness 24/7 worldwide. During the preaward activities, the DCMA provides precontractual advice to customers to help them construct sound solicitations, identify potential performance risks, select capable contractors, and write contracts that can be effectively administered. During contract administration, the DCMA assesses the contractors' business and technical systems to ensure their products, costs, and schedules comply with the terms and conditions of their contracts. The DCMA monitors contractor performance through data tracking and analysis, on-site surveillance, and tailored support to the program managers (data available online at http://www.dcma.mil/communicator/files/DCMA_Fact_Sheet.pdf.

MEASURING CONTRACTOR PERFORMANCE. As already stated, the contract-administration phase reflects the monitoring and controlling aspect of the PM processes. The monitoring and controlling project-management processes are focused on ensuring that project objectives are met as the PM performs such activities as measuring progress against plan, holding status meetings, and correcting the divergences from schedule or budget. A major emphasis of contract-administration activities is measuring contractor performance. The contractor's performance is measured to ensure that the actual contractor work results meet the cost, schedule, and performance standards agreed to in the contract. One of the key tools used in contract administration is earned value management. Earned value management is an integrated management approach to measuring a project's cost, schedule, and performance progress. This approach is based on comparing actual cost, schedule, and performance results with planned estimates. In earned value management, performance is measured by determining the budgeted cost of work performed and comparing it to the actual costs of the work performed. Thus, the project's progress is measured by comparing the earned value to the planned value ([11], p. 360). Earned value management is further discussed in Chapter 12 of this book.

MANAGE THE CONTRACT-CHANGE PROCESS. It would be nice to think that once the contract is awarded, the contract will never need to be changed or modified during the contract period. However, this is rarely the case. Contracts frequently require changes because of various reasons during the period of performance. These contract changes could result from a change in administrative requirements, such as a change in paying office or funding data, or any other administration issues. Contract changes could also result from significant revisions to the contract SOW, specifications, or other contractual requirements documents. A major part of contract-administration activities is focused on managing the contract-changes process.

Any changes to a contract are executed through a formal contract modification. *FAR* Part 43, Contract Modifications, prescribes policies and procedures for processing modifications to existing contracts. The *FAR* identifies two types of contract modifications: bilateral modification and unilateral modification.

A *bilateral modification* is a modification that is signed by the contractor and the contracting officer. Bilateral modifications are used to make changes to contracts that require agreement from both parties. These types of changes include negotiated equitable adjustments resulting from the issuance of a change order; definitized letter contracts; and other agreements of the parties modifying the terms of contracts ([2], Part 43.103).

A *unilateral modification* does not need to be agreed to by both parties; thus, it is signed only by the government contracting officer. These types of modifications are used to make administrative changers, issue change orders, make changes authorized by other *FAR* clauses, and issue termination notices ([2], Part 43.103).

A contract change order is a type of unilateral contract change that is issued pursuant to the changes clause of the contract. The change order allows the government the flexibility to make unilateral changes in designated areas within the general scope of the contract ([2], Part 43.201). A key component to the government contract-changes process is the use of the changes clause contained in the contract. The changes clause is a required clause for most government contracts. The changes clause allows the government contracting officer to make unilateral changes within the general scope of the contract to the contract drawings, designs, or specifications, or to the method of shipment or packing or to the place of delivery ([2], Part 52.243-1). A critical aspect in the use of the changes clause is that the direct changes must be within the scope of the contract. The scope of the contract reflects the work that was reasonably contemplated by the parties at the time of initial contract award. Proposed changes to the contract that are determined to be outside the scope of the contract are not permitted to be executed using the changes clause. Because these changes are determined to be out of scope, they must be executed through a new procurement action.

Controlling Contract Changes. It is paramount that both parties to the contract maintain a formal, disciplined, and methodical process for managing contract changes. Both parties, the buyer and seller, should ensure that only authorized individuals can initiate, process, and approve contract changes. In addition, it is essential that all changes to the contract are documented and processed as a formal modification to the contract document. A best practice in contract administration is the establishment of a contract change board (CCB) that will review and approve all major changes to the contract. This contract change board should evaluate every proposed major contract change to identify its potential impact on the contract cost, schedule, and performance baseline prior to approving and executing the contract change. The CCB is typically led by the program manager or other senior program-management official. Key members of the CCB include the contracting officer, financial manager, and lead technical manager.

Manage the Contractor Payment Process. An important part of the contract-administration phase is managing the contractor payment process. The method of contractor payment typically depends on the type of contract (fixed-price or cost-reimbursement) and the period of performance. For firm-fixed-price contracts, the contract may involve a single payment of the contract price upon final delivery and acceptance of the supply or service. Fixed-priced contracts with longer periods of performance may have progress payments, either based on cost or percentage of completion of work. Cost-reimbursement contracts are typically paid using monthy invoices.

10.4.6 CONTRACT CLOSEOUT/TERMINATION

Typically, a government contract can end in one of three ways: first, the contract can be successfully completed; second, the contract can be terminated for the convenience of the government; or third, the contract can be terminated for default. Regardless of how the contract ends, in the end all contracts must be closed out. Thus, the final phase of the contracting process is the contract closeout/termination phase.

CONTRACT CLOSEOUT. This is the process of "verifying that all administrative matters are concluded on a contract that is otherwise physically complete" ([5], p. 185). This process includes activities such as the processing of government property dispositions, final acceptance of products or services, final contractor payments, and documentation of contractor's final past-performance report.

The closeout of contracts that are physically complete requires the verification of documentation that reflects the completion of all required contractual actions. A contract is considered to be physically completed when the contractor has completed the required deliveries and the government has inspected and accepted the supplies; the contractor has performed all services and the government has accepted the services; and all option provisions have expired. In addition, the contract is considered to be physically completed when the government has given the contractor a notice of complete contract termination ([2], Part 4.804-4).

Once the contracting officer receives a contract completion statement from the contract-administration office, the contracting officer initiates the contract closeout process. *FAR* Part 4, Administrative Matters (and specifically *FAR* 4.804, Closeout of Contract Files), provides the specific policy and guidance for contract closeout. Furthermore, Department of Defense Form 1597, Contract Closeout Checklist, is used to ensure that all required contract actions have been satisfactorily accomplished.

CONTRACT TERMINATIONS. In he event that contracts are not successfully completed, contracts can be terminated through a termination-for-convenience or a termination-for-default process.

Termination for convenience is the "exercise of the government's right to completely or partially terminate performance of work under a contract when it is in the government's interest" ([2], Part 2.101). The federal government, as a sovereign entity, has the unilateral right to terminate any contract for its convenience, without any prejudice to the contractor. *FAR* Part 49 provides policy and guidance for terminations for convenience. Typically, contracts terminated for convenience provide a settlement compensating the contractor for the work done and might also include a reasonable allowance for profit.

Termination for default is the "exercise of the government's right to completely or partially terminate a contract because of the contractor's actual or anticipated failure to perform its contractual obligations" ([2], Part 2.101). A contractor terminated for default might be liable to the government for cost of excess procurement. In addition, a termination for default can result in a negative past-performance assessment for the contractor.

The termination for default procedure might involve the issuance of a cure notice and/or a show-cause notice. A *cure notice* is a written notice sent by contracting officer to the contractor that is failing to meet contract requirements. The notice requires a contractor to fix, or "cure," the contract deficiency within a given time period. A *show-cause notice* is a written notice sent by contracting officer notifying the contractor of the possibility of termination. The notice requests the contractor to "show cause" why the contract should not be terminated for default ([2], Part 49). *FAR* Part 49 provides policy and guidance relating to complete or partial termination of contracts for the convenience of the government or for default. Box 10.6 illustrates the use of termination for convenience in the Navy Littoral Combat Ship Program.

BOX 10.6 TERMINATION FOR CONVENIENCE

In April 2007, the Secretary of the Navy announced that the Department of the Navy was terminating construction of the third Littoral Combat Ship (LCS 3) for convenience under the termination clause of the contract because the Navy and the contractor could not reach agreement on the terms of a modified contract. The Navy issued a stop-work order on construction on LCS 3 in January 2007, following a series of cost overruns on LCS 1 and a projection of cost increases on LCS 3, which are being built under a cost-plus contract. The Navy announced in March 2007 that it would consider lifting the stop-work order on LCS 3 if the Navy and the contractor could agree on the terms for a fixed-price incentive agreement by mid-April. The Navy worked closely with the contractor to try to restructure the agreement for LCS-3 to more equitably balance cost and risk, but could not come to terms and conditions that were acceptable to both parties [12].

10.5 CONTRACT MANAGEMENT FROM THE CONTRACTOR'S PERSPECTIVE

So far, this chapter has discussed the contract-management process from the perspective of the buying organization. In discussing the contract-management process from the buyer's perspective, we presented the six phases of the contract-management process—procurement planning, solicitation planning, solicitation, source selection, contract administration, and contract closeout. Because contract management reflects the processes used by both buyers and sellers in managing contracts, this chapter would be incomplete without some discussion on contract management from the seller's perspective.

Just as the buyer performs specific activities during the contract-management process, so does the seller have specific processes and activities in performing contract management. Garrett identifies these seller contract management processes as presales activity, bid/no-bid decision making, bid/proposal preparation, contract negotiation and formation, contract administration, and contract closeout/termination as illustrated in Fig. 10.5 [5]. These seller-specific contract-management processes and activities will now be briefly presented.

While the buying organization is performing the preaward activities of procurement planning, solicitation planning, and solicitation, sellers (prospective contractors) are also performing preaward activities, including presales activity, bid/no-bid decision making, and bid or proposal preparation.

10.5.1 *Presales Activity*

Most defense contractors are very proactive with prospective and current defense buying organizations. As part of their business-development effort, these companies conduct presales activities such as identifying prospective and current customers, determining customers' current needs and future requirements, and evaluating competitors. The presales activity phase of the seller's contract-management process is focused on conducting market research in its specific defense industry sector, benchmarking products and

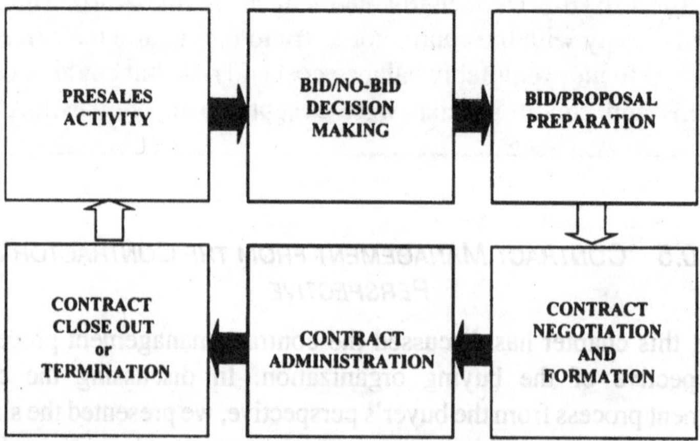

Fig. 10.5 Contract-management process (seller's perspective).

processes, and conducting competitive analysis as methods to improve customer focus and to increase its competitive advantage [5].

10.5.2 BID/NO-BID DECISION MAKING

After the buying organization has completed the solicitation phase of the contract-management process and the solicitation has been issued, the selling organizations then initiate the bid/no-bid decision-making process. This process entails the seller performing a risk-and-opportunity assessment against the buyer's solicitation. Using tools such as SWOT (strengths, weaknesses, opportunities, threats) analysis and risk-management activities, the seller can make a risk-based determination whether or not to prepare a bid for the specific solicitation [5].

10.5.3 BID/PROPOSAL PREPARATION

If the outcome of the bid/no-bid decision-making process is a decision to pursue the specific government contracting opportunity, the bid/proposal preparation phase begins. This phase includes the process of developing offers in response to a buyer's solicitation (or, in the case of unsolicited proposals, based on perceived buyer needs) for the purpose of persuading the buyer to enter into a contract [5].

10.5.4 CONTRACT NEGOTIATION/FORMATION

As the buying organization progresses through the source-selection phase of the contract-management process, the selling organization is progressing through the mirror-image phase of source selection, that is, the contract negotiation and formation phase. In this phase, the prospective contractor is progressing towards reaching an understanding of the nature of the project and negotiating the contract terms and conditions with the buying organization for the purpose of developing a set of shared expectations and understandings [5]. Of course, it should be noted also that during the contract negotiation and formation phase, if the prospective contractor cannot reach agreement with the buying organization on the terms and conditions, or any other aspect of the contract, it may be in the best interest to walk away from the negotiations and the proposed contract altogether. Prospective contractors should "not agree to bad deal." Even in defense acquisition contracts, "no business is better than bad business." ([5], p. 148). Box 10.6 also illustrates this point.

10.5.5 CONTRACT ADMINISTRATION AND CONTRACT CLOSEOUT/TERMINATION

Once the contract is awarded, both the buyer and the seller have an established and agreed-to contractual baseline in terms of cost, schedule, and performance requirements of the contract. Because each party has an identical copy of the contract, both are now on the same sheet of music, so to speak; each can now begin performing its respective postaward activities within the contract-administration phase and eventually contract closeout phase.

The contract-administration phase (which involves a process of ensuring that each party's performance meets contractual requirements) is identical for both the buyer and seller. Both parties to the contract should have formal, disciplined, methodical processes for conducting the various contract-administration activities. As the buyer conducts preperformance conferences, monitors the contractor's performance, measures the contractor's work, and manages the contract-change process, the contractor is attending the preperformance conferences, performing the contractual effort, and either responding to government-initiated contract changes or initiating contractor-proposed contract changes.

Eventually, the contract will be successfully completed. Or, the government, depending on the government's needs or the contractor's performance, will decide to terminate the contract for its convenience or for the default of the contractor. Regardless of how the contract ends, all contracts must be closed out, and both the government and contractor will perform the contract closeout process and related activities.

10.6 EMERGING TRENDS IN DEFENSE CONTRACT MANAGEMENT

Contract management in the Department of Defense continues to be a critical aspect of defense acquisition management, as well as a significant contributor to the national military strategy. As the DoD realizes the significance of irregular warfare in the pursuit of the national military strategy, the importance of procuring defense weapon systems and services meeting time, cost, and performance requirements will continue to increase.

Although contract management is crucial to the DoD's mission, historically, the Department has not been successful in managing its contracts for major defense systems and services. The Government Accountability Office (GAO) continues to list DoD contract management as a high-risk area. Indeed, it has designated DoD contract management as a high-risk activity since 1992. The GAO cites poor acquisition outcomes and missed opportunities to improve its approach to buying goods and services as the reason for the high-risk rating [13].

The current dynamics of contract management in the DoD reflect the following emerging trends:

1) *Increasing budgets*: In FY 2006, DoD procurement spending reached approximately $305 billion [14]. The DoD-proposed FY 2008 budget, including supplemental funding for the wars in Iraq and Afghanistan, is over $624 billion—the highest it has been since 1946 [15].

2) *Decreasing workforce*: The pending departure of the Baby Boomer generation contracting workforce will require the DoD to continue to take action to recruit, retain, and train the contracting workforce. The Baby Boomer generation comprises 71% of the DoD workforce and 76% of the acquisition, technology, and logistics (AT&L) civilian workforce. The DoD contracting workforce includes 27,742 members and represents 22% of the DoD AT&L workforce (data available online at http://www.dau.mil/spotlight/doc/Final%20Final%20Report.pdf.

3) *Consolidated defense industry*: The defense industry is dominated by a handful of major contractors consisting of Lockheed Martin Corporation, Boeing Company, Northrop Grumman Corporation, General Dynamics, and Raytheon Company [16]. The companies do not only receive large sums of money in terms of contract awards, but they can also perform program-management functions as lead systems integrators on many major defense contracts.

4) *Timely delivery, less development*: The DoD is shifting away from high-dollar weapons programs with long development and production timelines. Instead, the Department recently prefers the use of more commercial, off-the-shelf systems that provide more timely deliveries meeting immediate warfighter needs [15].

5) *Increase in contracts for services*: The services acquisition volume in the DoD has continued to increase in scope and dollars in the past decade. The DoD has spent more on services than on supplies, equipment, and goods, even considering the high value of weapon systems and large military items [17]. Between FY1999 to FY2003, the DoD's spending on services increased by 66%, and in FY2003 the DoD spent over $118 billion (or approximately 57% of total DoD procurement dollars) on services [18]. The acquired services presently cover a very broad set of service activities, including professional, administrative, and management support; construction, repair, and maintenance of facilities and equipment; information technology; research and development, and medical care. As the DoD's services acquisition volume continues to increase in scope and dollars, the agency must give greater attention to proper acquisition planning, adequate requirements definition, sufficient price evaluation, and proper contractor oversight [19].

10.7 CONCLUSION

This chapter provided an overview of the contract-management processes, concepts, and terms that are used by defense acquisition managers, defense organizations, and defense contractors in managing defense acquisition programs. The focus was on the pre-award and post-award contract-management processes as performed by both government and contractor organizations. A discussion of basic government contract concepts was first presented, with a review of sources of contracting authority, contracting regulations, statutory requirements, and some common government socioeconomic programs. The roles and responsibilities of the contracting officer, PM, and contractor personnel were then discussed, as well as the relationship between the contracting officer and other members of the project team. The contracts management process was presented, focusing on the common phases and activities that occur during the process from both the government and contractor perspectives. Finally, emerging trends in defense contracting were presented, along with highlights of current challenges and opportunities. One of the goals of this chapter was to present contract management as "the language of defense acquisition" and as an essential part of the management of defense acquisition projects.

REFERENCES

[1] Meigs, W. B., Johnson, C. E., and Meigs, R. F., *Accounting: The Basis for Business Decisions,* 4th ed., McGraw-Hill, New York, 1977, p. 91.

[2] *Federal Acquisition Regulation (FAR),* Washington, D.C., www.arnet.gov/far [retrieved Nov., 2007].

[3] Forsberg, K., Mooz, H., and Cotterman, H., *Visualizing Project Management*, Wiley, Hoboken, NJ, 2005, p. 1.

[4] Hirsch, W. J., *The Contracts Management Deskbook,* rev. ed., American Management Association, New York, 1986, p. 4.

[5] Garrett, G. A., *World Class Contracting,* CCH, Chicago, 2007, pp. 80–194, 386–409.

[6] Office of the Inspector General, "Procurement Policy for Armored Vehicles," DoD IG Report No. D-2007-107, U.S. Dept. of Defense, Washington, D.C., 27 June 2007.

[7] Defense Acquisition Univ., "Glossary of Defense Acquisition Acronyms and Terms," Fort Belvoir, VA, July 2005.

[8] Office of the Inspector General, "Air Force KC-X Aerial Refueling Tanker Aircraft Program," Rept. No. D-2007-103, U.S. Dept. of Defense, Washington, D.C., 30 May 2007.

[9] Government Accountability Office, "Defense Acquisitions: DOD Has Paid Billions in Award and Incentive Fees Regardless of Acquisition Outcomes," GAO-06-66, Washington, D.C., Dec. 2005.

[10] Government Accountability Office, "Defense Acquisitions: DOD Wastes Billions Of Dollars Through Poorly Structured Incentives," GAO-06-409T, Washington, D.C., 5 April 2006.

[11] Project Management Inst., *A Guide to the Project Management Body of Knowledge (PMBOK Guide),* Newtown Square, PA, 2004, pp. 269–297.

[12] Office of the Assistant Secretary of Defense (Public Affairs) News Release, No. 422-07, U.S. Dept. of Defense, Washington, D.C., 12 April 2007.

[13] Government Accountability Office, "High-Risk Series: An Update," GAO-07-310, Washington, D.C., Jan. 2007.

[14] Clark, T., "Procurement Accounts," *Government Executive,* 6, Aug. 2007, p. 6.

[15] Grant, G., "Defense Dollars," *Government Executive,* 50, Aug. 2007, p. 50.

[16] Shoop, T., "Onward and Upward," *Government Executive,* 39, Aug. 2007, p. 39.

[17] Camm, F., Blickstein, I., and Venzor, J., *Recent Large Service Acquisitions in the Department of Defense,* RAND National Defense Research Inst., Santa Monica, CA, 2004, pp. 1–5.

[18] Government Accountability Office, "Contract Management: Opportunities to Improve Surveillance on Department of Defense Contracts," GAO-05-274, Washington, D.C., March 2005.

[19] Government Accountability Office, "Best Practices: Taking a Strategic Approach Could Improve DOD's Acquisition of Services," GAO-02-230, Washington, D.C., Jan. 2002.

STUDY QUESTIONS

10.1 Comment on the statement "contracting is the language of defense acquisition."

10.2 How is contract management an essential part of the management of defense acquisition projects?

10.3 Identify the various sources of contracting authority for government contracting officers.

10.4 What are the major contracting regulations used in defense contracting?

10.5 What are some of the statutory requirements and some common government socioeconomic programs that are supported by the defense contracting process?

10.6 What are the roles and responsibilities of the contracting officer, program manager, and contractor personnel. In addition, what is the relationship between the contracting officer and other members of the project team?

10.7 What are the six phases of the contract-management process, from both the government and contractor perspectives?

10.8 What are some of the contracting activities involved in each of the six phases of the contract-management process from the government perspective?

10.9 Differentiate between product/service descriptions and specifications.

10.10 Explain the difference between a statement of objective and a statement of work.

10.11 What is the purpose of market research?

10.12 Differentiate between sealed bidding and negotiated procurement procedures.

10.13 What are the exceptions to the competition requirements of the Competition in Contracting Act?

10.14 Differentiate between fixed-price and cost-reimbursement contracts.

10.15 What is the Government Point of Entry?

10.16 Differentiate between presolicitation, preproposal, and preperformance conferences.

10.17 Discuss the best-value continuum.

10.18 Describe the source-selection organization.

10.19 What is the purpose of the contract-administration phase?

10.20 Discuss the difference between bilateral and unilateral contract modifications.

10.21 How important is controlling contract changes?

10.22 Discuss the difference between termination for default and termination for convenience.

10.23 What is the purpose of contract closeout?

10.24 What are some of the contracting activities involved in each of the six phases of the contract management process from the contractor perspective?

10.25 Describe some of the emerging trends and current challenges and opportunities in defense contracting.

Chapter 11

Financial Management

PHILIP J. CANDREVA*
NAVAL POSTGRADUATE SCHOOL, MONTEREY, CALIFORNIA

Learning Objectives

- Describe the DoD financial management framework and how it overlaps with the acquisition framework, including roles and functions and the DoD resource-allocation process
- Explain the characteristics of a "good" budget for an acquisition program
- Give an overview of the key concepts of federal appropriations law that affects acquisition
- Discuss the roles of, and the program office's relationship with, Congress and industry with respect to the financial aspects of acquisition management
- Identify how key financial-management concerns evolve along an acquisition program's life cycle
- State how timeless questions in public financial management shape the environment for defense acquisition
- Apply concepts from this chapter to real-world examples

11.1 Introduction

This chapter focuses on the financial aspects of project management in the Defense Department. In the middle of the first decade of the new century, when most of the federal government's attention on military affairs was devoted to the war on terrorism and Iraq, several key reports commented on

*Senior Lecturer of Budgeting and Senior Associate with the Center for Defense Management Reform, Graduate School of Business and Public Policy.

the difficulties of managing defense acquisition, some of which related to financial management:

> Virtually all program managers we spoke with first defined success in terms of enabling warfighters and doing so in a timely and cost-efficient manner. But when the point was pursued further, it became clear that the implied definition for success in DOD is attracting funds for new programs, and keeping funds for ongoing programs ([1], p. 56).

> Our assessment concluded that [...] Current budget reallocations, and or, shortfalls are frequently resolved by stretching programs, thereby introducing instability and long-term cost growth. In taking these actions, the Department accepts long-term cost increases and delays in acquisition programs to achieve short-term savings and budget flexibility ([2], p. 32).

> Cost overruns, schedule slips, and performance shortfalls have plagued large weapon system acquisition programs since World War II [...] identified a range of contributing factors to these problems including requirement and funding instability ([3], p. 20).

And that constructive criticism has not fallen on deaf ears. The defense-acquisition community and the Business Transformation Agency have given these recommendations serious attention:

> The Department is striving to budget programs realistically through the Planning, Programming, Budgeting and Execution process. [...W]e are implementing several approaches that substantially will improve the rigor and focus of our requirements development, budgeting, acquisition and sustainment planning processes ([4], p. 20).

The first section of this chapter provides a foundational look at the DoD financial-management framework and how that framework supports, as well as constrains, project management. It describes the roles played by various financial-management actors and the critical processes those actors support. These are the roles and processes that affect the success of a program. The second and third segments cover the planning, programming, budgeting, and execution (PPBE) system—emphasizing the budgeting and execution phases, respectively. This section will explain the characteristics of a "good" budget, along with the legal and policy limitations on the use of the funds provided by that budget. The successful acquisition manager must be proficient at both acquiring financial resources and using them effectively, all within legal bounds. The fourth part of the chapter looks outside DoD to the project's relationship with Congress and industry. The former provides the spending authority, and the latter receives much of it. Creating a feasible plan is necessary, but insufficient, to successfully implement it; the program manager must be able to anticipate concerns and work well with these critical stakeholders. Lastly, this chapter considers timeless questions related to defense financial management and how they partially define the environment for project managers.

11.2 DoD Financial-Management Framework

Financial management in the DoD is not a simple concept. Although most defense project managers are concerned with acquiring and managing funds in support of their project objectives, surrounding them are processes and actors with much broader concerns. As part of the federal government, the DoD is heavily influenced by a larger fiscal environment that ranges from the fiscal policy that drives macroeconomic decisions and the defense budget top line to the esoteric world of federal accounting standards and computing daily cash balances at the U.S. Treasury.

Roles and functions form part of this framework. Those roles and functions are engaged in several acquisition-related processes and produce myriad products, including strategies and plans, decision-support assessments, requests for resources (budgets), reports of progress, and management-control activities. The different military services and even different commands within the same service might perform financially oriented roles at different places within the organization, but each of the components perform all of the same basic roles. Each of these roles will be discussed next within an organizational model familiar to the author. The student should not interpret this as the only or best way to arrange these functions, but rather as an illustrative example. Although all of these roles contribute in some way to defense financial management and might impact an acquisition program manager, unfortunately they sometimes "are perceived as separate players" ([5], p. 9) and might not function as a highly integrated community, thereby negatively impacting the quality of information and support.

11.2.1 Program Manager

The program manager, of course, has overall responsibility and accountability for achieving the program's objectives. As we saw in this chapter's introduction, this requires a significant amount of attention to fiscal matters—obtaining and preserving resources sufficient to execute the plan or articulating revised plans if budgets fall short of desires. The program manager is supported by a broad cast of other actors.

11.2.2 Business and Financial Manager

As a member of the program team, the business and financial manager (BFM) has primary responsibility for financial matters in the program office. These responsibilities broadly range from acquiring funds through the building and defending of budgets to the execution of those funds. To do this job effectively, the BFM should maintain insight on all key programmatic matters and be both technically proficient and politically savvy. The BFM's broad perspective often makes him or her a key advisor to the program manager. The BFM role exists not only at the program level, but often also exists at the Program Executive Officer (PEO) level, having supervisory and support responsibility for subordinate program BFMs.

11.2.3 CONTRACTING OFFICER

A large percentage of an acquisition program's budget goes to contractors; therefore, the awarding and administration of those contracts is a critical function. The contracting officer's role includes contracting strategies, designing contracts, creating incentives, and monitoring tasks to ensure that both parties meet their obligations, that goods and services are received, accepted, and paid for, and that any disputes are settled. The contracting officer is the program manager's agent, and he/she also ensures compliance with myriad federal contracting rules.

11.2.4 COMPTROLLER

Whereas the program office has the detailed information and the BFM does the majority of the work building the budget, the supporting comptroller has the overall responsibility for its preparation, compilation, and submission. Comptrollers serve a gatekeeper role, establishing procedures, enforcing standards, and ensuring compliance with policy. They will track budgets through the service, Office of the Secretary of Defense (OSD), and congressional review processes. During execution, comptrollers serve a different gatekeeper function, focused on compliance with fiscal law and the provisions of the budget and its associated appropriations. They will ensure adequate internal controls exist over fiscal processes; they perform the accounting functions, coordinate with Defense Finance and Accounting Service (DFAS), service headquarters financial management organizations, and conduct mid-year and end-of-year financial performance reviews. Whereas the BFM is the advocate for the program, the comptroller advocates for compliance; whereas the BFM might have an incentive to be aggressive, the comptroller has an incentive to be conservative. This tension is healthy and will often result in the program pushing against the myriad rules in the advancement of its objectives, while being restrained from actually breaking the rules.

11.2.5 RESOURCE SPONSOR

In the PPBE process, certain offices on the service headquarters staff are often referred to as resource sponsors because they are responsible for ensuring there is an adequate stream of resources to their assigned functional area. For example, naval aviation has a resource sponsor for tactical aircraft and another for support aircraft; each advocates for sufficient resources for his/her portion of the naval program. The sponsors craft a portfolio of programs that provide the greatest military utility for the available dollars. To be effective, both the resource sponsors and their affiliated program offices maintain active communication and insight. The resource sponsor needs detailed information about program capabilities and near real-time status; likewise, the program needs policy direction and advocacy in the Pentagon.

11.2.6 ACCOUNTING AND BILL PAYING

Although sometimes viewed as "back-office" functions, accounting and bill paying are an inherent part of DoD financial management. Many of these tasks are performed on behalf of the military services by the DFAS, with the services providing input ranging from the issuance of funding documents and other evidence of obligations to certified invoices for payment. Accountants within the services input transactional data on the movement of funds, obligations, accounts payable, and the like, whereas the DFAS prepares the reports and holds the accounts of record. The program office, in cooperation with its servicing comptroller, must ensure that its memorandum records of the financial status of the program match the official records of the DFAS. This complexity of defense appropriations and fiscal law demands a profound amount of reconciliation and cross-checking to keep all of the accounts straight.

11.2.7 COST ANALYST

Budgets and spending are not the same as cost. Budgets are plans and formal requests for resources. The congressional authorization and appropriation that result provide a fraction of the resources necessary to carry out the program's objectives, generally a fraction equal to one year's worth of the overall program's needs. Spending is a tabulation of the resources actually used in a given period of time. Budgets and appropriations are the focus of the BFM, comptroller, resource sponsor, and the DFAS, yet others are concerned about cost. Cost is different. Cost is normally not considered on a time basis, like budgets and appropriations and spending; rather, cost is calculated on a product or process basis. Cost analysis supports the planning and programming process from the perspective of the long-term affordability of military capability. It supports life-cycle-management decisions, and it supports the negotiation of competitive contracts. Each service has an independent cost-estimating team that takes on this role and that serves as a check and balance on the overall resource-allocation/decision-making processes.

11.2.8 AUDITOR

Whereas many of the roles just described result in actions, the role of auditors in the DoD is to review actions already taken. The objective of the audit function is to evaluate program performance; compliance with law or policy, adequacy of control mechanisms, or the integrity of financial reporting or data systems. The purpose of the evaluation is to ensure program performance; that information is timely, relevant, and reliable; and that resources are safeguarded and used properly (in conformance with laws and policies), as well as economically.

11.3 BUDGETING

"Simply stated, a budget is a quantified, planned course of action over a definitive time period. It is an attempt to estimate inputs and the costs of inputs along with associated outputs and revenues from outputs" [6]. The Navy's *Budget Guidance Manual* states, "A budget is defined as a document that expresses in financial terms the plan for accomplishing an organization's objectives for a specified period of time. It is an instrument of planning, performance measurement, decision-making, and management control, as well as a statement of priorities" ([7], part 1, p. I-2). Both definitions contain similar phrasing: a plan for achieving certain objectives, a time horizon, and financial resource requirements. The process for building budgets for acquisition programs is the focus of this section.

11.3.1 PROCESS OVERVIEW

The DoD develops its annual budget using the PPBE system. Variants of this system have existed since the early 1960s. The description here intends to be as general and timeless as possible. The B in PPBE stands for "budgeting," but one cannot get to the B without first going through the Ps. In planning, the OSD and Joint Staff, with input from the unified and combatant commands, set the strategic direction and priorities for the department. The Strategic Planning Guidance (SPG), issued by the Secretary of Defense, articulates national security strategies, threat assessments, and fiscally informed guidance to the services regarding requirements for force structure, readiness, supporting infrastructure, and modernization. It supports the latest versions of the National Security Strategy, National Military Strategy, and *Quadrennial Defense Review*. The product from the planning phase, which defines the start of the programming phase, is the Joint Programming Guidance (JPG). Whereas the SPG was *fiscally informed*, the JPG is *fiscally constrained*. The services and defense agencies are given financial targets and guidance for developing their program objectives memoranda (POM).

The POM is a portfolio of specific programmatic information within that resource constraint, which represents what the component believes will best meet the direction in the JPG. The POM further contains information about gaps between what is affordable and what is desirable. Not all objectives of the JPG can be satisfied within the resource constraint, and those risks are articulated in the POM. In practice, the POM can be developed in different ways. The U.S. Marine Corps tends to create an "order of buy" by developing a weighted prioritized list of programs ranked according to a cost/benefit analysis. The U.S. Navy does a capability assessment by functional areas, integrates that across the service, and then determines how individual programs must be shaped to achieve those desired capabilities. Along with its capabilities and programmatic proposals, the Navy conducts pricing validations to ensure programs are resourced parsimoniously. Similarly, the U.S. Air Force uses a capability-based planning system that ranks programs by "most dear" to "least dear," and a line is drawn where resource limitations demand. Component POMs are submitted to OSD for review by the Secretariat staff and the Joint Staff. Final decisions reside with the Secretary of Defense and are recorded in program decision memoranda (PDM) (see Fig. 11.1).

The process is heavily calendar driven, with the first Monday in February as the "drop-dead" date for the submission of the President's budget to Congress. Moving backwards chronologically from there, the budget was

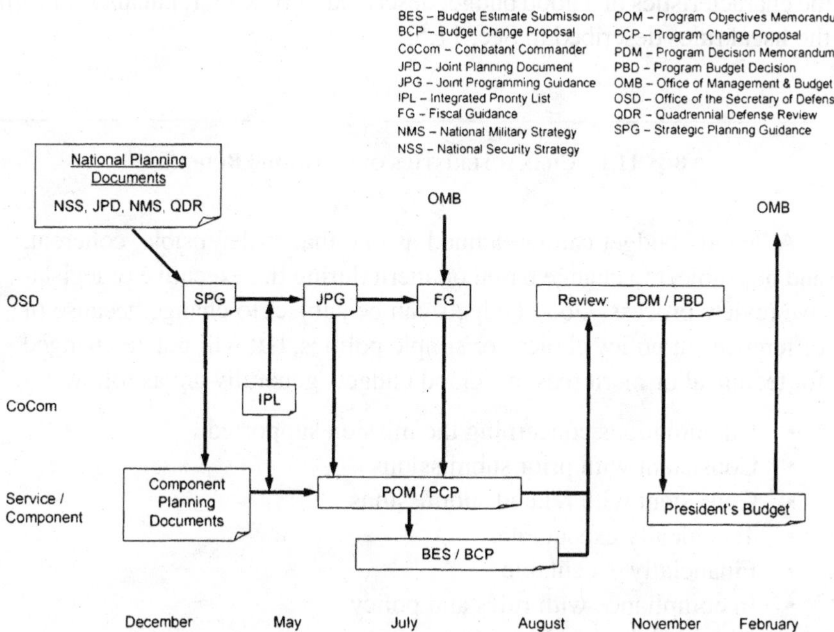

Fig. 11.1 Planning, programming, and budgeting process.

written in January reflecting the results of the OSD/OMB review in the fall. The components built their budgets in the summer, and initial programming decisions occurred in the spring. Because of the lead times and complexity of the budget, each phase actually begins before the preceding one ends. Thus, programming decisions are still being made well into the budgeting process, requiring rework and vigilance.

The budget, in essence, converts the POM into a prescribed format dictated by the OMB and Congress. The budget consists of justification material (also known as "j books" or "justifications of estimates") arranged by appropriation title and, in some cases, specific inputs or outputs. Examples of appropriation titles are Aircraft Procurement, Navy (APN) or Research, Development, Test and Evaluation, Army (RDT&E,A). Specific inputs and outputs can include information technology, petroleum products, or facilities modernization. For acquisition programs, the two critical exhibits are, first, the P-1, procurement line item, and its associated components, the cost analysis (P-5), procurement history and planning (P-5a), and production schedule (P-21). For established programs, a modification program (P-3a) can be included. The second critical exhibit is the R-1, RDT&E programs, and the individual item justification (R-2). All budget exhibits contain two significant parts: the narrative description and the financial information. A well-written narrative will not only describe what the funds are buying (outputs), but why they are being bought (the outcomes affected). This financial information, in general, must conform to the characteristics of a good budget described in Box 11.1, but also conform to the fiscal rules described in Sec. 11.5.

Box 11.1 Characteristics of a "Good" Budget

A "good" budget can be defined as one that is defensible, coherent, and not subject to change based on merit during the executive or legislative review process. Good budgets can be subject to change because of differences in policy choices or simple politics, but will not be changed for technical or merit reasons. Good budgets generally are as follows:

- Unambiguous concerning the mission supported
- Consistent with prior submissions
- Consistent with related submissions
- Technically executable
- Financially executable
- In compliance with rules and policy
- Able to delineate cost drivers
- Administratively correct

11.3.2 BUDGETING FOR ACQUISITION

Requirements exceed available resources, and competition is fierce for those constrained resources. This tension results in pressures to cut budget slack from programs. Too often program budgets can be reduced to the point that their schedule and performance criteria are infeasible at the level of resources assigned. Vigilance is necessary to ensure even tentative plans are executable—or at least that the level of risk is well understood. How do the PM and BFM generate defensible information to ensure they emerge with an adequate budget? This is where those supporting actors make their financial mark. As the POM is iteratively developed, and "what if" scenarios are debated, the PM and resource sponsor should be armed with estimates of budgetary requirements under different scenarios. Cost analysts should develop a set of estimates for likely scenarios: incremental increases or decreases to quantities being procured or accelerating or decelerating a development timeline. Contractors should be armed with cost-volume-price analyses that can help support budgetary estimates. Likewise, learning-curve estimates will be adjusted to fit various scenarios.

A starting point is the program baseline: program objectives, quantities, schedules, the detailed work breakdown structure, and the costs associated with those tasks. Once a schedule is developed, associated costs are categorized according to budget and appropriation policies. The program must determine whether costs are investments or expenses, RDT&E or procurement or Operation and Maintenance (O&M), fully or incrementally funded. It must also determine the correct fiscal year with which each cost is associated. This can be confounded by authorities for advance and/or multiyear procurement. (These concepts will be developed in the next section on funds execution.) The costs are often computed in today's dollars without regard to inflation; however, budgets are requests for future dollars, and so the costs must be accelerated according to inflation factors.

Budgets are requests for spending authority, and because budget authority is constrained by appropriation characteristics, the program office must anticipate its needs well in advance and request the right combination of funds to support program plans. As the program matures, the colors of money change, shifting primarily from research and development to procurement to operations and maintenance. The nature of the tasks, when they will occur, and how much they cost (as well as whether the program is financed incrementally or fully) all determine which funds should be used. In the early stages of concept refinement, RDT&E funds are used and will probably not have been appropriated for this particular effort. They likely are funds provided for broad categories of research. As ideas and technology mature, a program begins to take shape, and the type of R&D money will change. Table 11.1 displays the various types of R&D money found in the DoD and their relationship to the phase in the acquisition life cycle.

TABLE 11.1 RDT&E FUNDING TYPES [8]

Budget activity	Research category	Title	Description	Relation to acquisition life cycle
01	6.1	Basic research	Directed toward greater knowledge or understanding without specific applications in mind; farsighted, high-payoff research that provides the basis for technological progress	Pre-Milestone A
02	6.2	Applied research	Directed toward understanding the means to meet a recognized and specific need; can be oriented, ultimately, toward the design, development, and improvement of prototypes and processes; includes studies, investigations, and nonsystem specific technology efforts	Pre-Milestone B (concept & technology development)
03	6.3a	Advanced technology development (ATD)	Development of subsystems arid components and efforts to integrate them into system prototypes for field experiments and tests in a simulated environment; includes concept and technology demonstrations of components and subsystems or system models	Pre-Milestone B (concept & technology demonstration). Goal is to move out of science and technology and into the acquisition process during the FYDP.
04	6.3b	Advanced component development & prototypes (ACD&P)	Evaluate integrated technologies and prototype systems in a realistic operating environment; expedite technology transition from the laboratory to operational use; prove technology maturity to reduce technical risk	Pre-Milestone B (advanced component development activities & technology demonstrations). Progression of program phases or production funding must be in the FYDP.
05	6.4	System development & demonstration (SDD)	Engineering and manufacturing development tasks aimed at full-rate production; mature system development, integration, and demonstration, live fire, and initial operational test & evaluation	Past Milestone B, pre-Milestone C. A logical progression of program phases and production funding must be in the FYDP and fully funded.
06	6.5	RDT&E management support	Sustain or modernize installations or operations for general RDT&E; Funds test ranges, construction, and maintenance of laboratories, operation, and maintenance of test aircraft and ships	N/A. Personnel are funded through one of the program areas, as applicable.
07	6.6	Operational system development	Development efforts to upgrade systems that have been fielded or have been approved for full-rate production	Past Milestone C. A logical progression of program phases and production funding must be in the FYDP and fully funded.

The purpose of procurement funds is to buy end items or those things associated with delivering end items. Normally, procurement funds are not used until the program is authorized to enter production. One of the financial challenges is to build a POM several years in advance and to refine that plan, budget two years in advance and hold to that plan, and be accountable as circumstances and conditions change. Having enough RDT&E funding to address the unforeseen difficulties while planning for and being able to spend the procurement money once it is appropriated are difficult problems to manage.

Just as RDT&E funding comes in more than one type depending on the nature of the work, procurement funds also come in several forms. For instance, Navy tactical aircraft are procured with APN funds. Budget activity 1 (APN-1) buys the airframe and engine; APN-6 procures initial spare parts; APN-7 acquires support equipment; and, once fielded, APN-5 procures modifications to the aircraft. Potentially, a program office might budget for up to four types of procurement funds. It is incumbent upon the program BFM to understand when each type is required in the program plan, and in what amount, and how it is affected by programmatic decisions.

As the General Accounting Office (GAO) recently put it, "If we expect programs to be executed within budget, programs need to begin with realistic budgets. The foundation of an executable budget is a realistic cost estimate that takes into account the true risk and uncertainty in a program" ([9], p. 17). Obviously, early in the system's life cycle there is little, if any, detailed information on which to base an estimate. Requirements might not be fully established, let alone a design or knowledge about manufacturing costs. Thus, estimating techniques early in the process produce more gross estimates and are more prone to error than the more detailed and refined estimates that come later. It is beyond the scope of this book to describe the details of cost estimating. However, the four methods most commonly used in defense and their place in the life cycle of the program are described next.

Early in the acquisition process, as concepts are still being explored, there are more options than decisions; a design does not yet exist; production processes have not yet been established. But there exist historical data about similar programs in the past. The *analogy* method uses historical data for similar systems, adjusted for known changes. Another cost-estimating method used early in the acquisition process is *parametric analysis*. Parametric analysis uses known relationships between variables and costs; for example, ship tonnage and aircraft weight are closely correlated to construction costs. Once a system design matures, *engineering* estimates break it down into its components, subsystems, assemblies, and parts. Each of those smaller units might have known costs if it is existing technology. If it is new technology, estimates can be based on material and labor costs. Each component's cost, plus the cost of integrating those components, can then be added together into larger and larger units; the process is largely "bottom up." Obviously, the most detailed and generally most accurate method of cost estimating is to use the system's own *actual costs*. In building budget estimates for production, one should incorporate learning curves, economies of scale, the allocation of overhead across a larger base, and other factors of scale. The astute student should realize that it is possible to mix these forms of estimating. The program office should employ the method that produces the estimate most likely to be accurate; the method, however, must be defendable under the scrutiny of the PPBE and acquisition management systems.

A major consideration in estimating costs for production phases of acquisition projects is the concept of learning curves (or experience curves). Learning curves reflect the reduction in labor hours (and associated costs) that occur as repetitive manufacturing and assembly tasks are accomplished in production. In its simplest form, learning theory states that *the labor required to produce an item decreases by a fixed percentage as the production quantity doubles*. For example, if 80 labor hours are required to produce the 100th item, and 60 hours are required to produce the 200th item, then the learning rate is 75%. Using this learning rate, an estimate of the labor hours to produce the 400th item would be 45 hours.

Learning in production tasks occurs for a variety of reasons, such as worker efficiency, standardization, and improvements in methods and processes. More learning typically occurs on tasks that involve large production quantities with high amounts of human "touch labor." Less learning occurs on low-volume production projects, as well as those that are highly automated. Thus, different commodities and different industries demonstrate different learning rates. For example, a project to produce only four satellites would likely have very little, if any learning, whereas a project to produce 1000 missiles would likely exhibit substantial learning.

Learning can be impeded for a variety of reasons. For example, breaks in the production line for a period of time might cause workers to "forget" the efficiency with which they had performed in the past. Design changes, changes in production processes, and worker turnover are among the other reasons that learning might not occur as expected.

Attention to learning rates is essential in estimating production costs for many types of defense projects. For those projects in which substantial learning can be expected, the reduction in labor costs over time can be significant. Thus, PMs and contractors both have incentives to promote manufacturing conditions under which learning occurs and is sustained.

In summary, budgeting is an articulation of the programmatic decisions in the POM in a format specified by the OMB and Congress that links program objectives, schedules, and financial resource requirements. The financial resource requirements are presented in particular ways based on the nature of the work being performed, how that work conforms to general fiscal law, and any specific authorities held by the program office. The next section discusses those laws and possible authorities. One should understand that the expectation of both the DoD leadership and the Congress is that funds will be requested in the form in which they are desired, that funds are appropriated in light of that request (but might not exactly match it), and that they are expected to be executed as they were appropriated (within the bounds of some inherent, but limited, flexibility).

11.4 FUNDS EXECUTION

Budgeting (requests for authority), appropriations (grants of authority), and execution (exercise of that authority) must be done consistently. Because the application of fiscal law occurs in the spending, or execution, of the funds, a brief legal discussion is presented here. However, it could very well have been presented in the preceding section on budgeting because budgets must request funds in accordance with these principles. This section of the chapter provides a very basic introduction to the key concepts of fiscal law that most commonly relate to acquisition. As in Sec. 11.3, general principles will be explained first, followed by their specific application to the world of defense acquisition.

11.4.1 LEGAL FRAMEWORK

In its most basic sense, fiscal law is concerned with the obligation and expenditure of federal funds. It could be said to begin with the enactment of an appropriation, which grants an agency the authority to obligate the government to make payments from the Treasury in support of policy objectives. The word *appropriation* originates from the Latin word *appropriare*, which means "to set aside" or "to make one's own." An inherent meaning of appropriation is that funds are designated for particular uses (set aside) to the exclusion of other uses (to make one's own). There are three critical concepts in the definition of an appropriation that will be explored in this section. First is the notion that funds are designated for particular uses and times. The second concept is contained in the phrase "to the exclusion of other uses." Appropriations are a positive authority, which means that the authority does not exist until enacted. The agency is not permitted to spend as it sees fit unless authorized by an appropriation—quite the opposite, in fact: an agency can only spend if and only if funds are appropriated. The third concept is that an appropriation provides budget authority, not money. The agency merely has permission, or authority, to put the government in a position to make a payment now or at some point in the future. The agency does not actually hold the funds; the Treasury does.

Appropriations are classified in several categories based on characteristics of *purpose*, *time*, and *amount*.

PURPOSE. The most fundamental statute dealing with the use of appropriated funds is 31 U.S. Code Section 1301(a): "Appropriations shall be applied only to the objects for which the appropriations were made except as otherwise provided by law" [10]. This statute prohibits funding authorized items with the wrong appropriation. It also prohibits any source of funding to be applied to unauthorized items. Generally, the language in an appropriations bill is fairly clear, but not always. There are some established guidelines:

- A specific appropriation must be used to the exclusion of a more general appropriation that might otherwise be viewed as available.
- Because appropriations are designated for particular uses to the exclusion of other uses, one cannot borrow funds from one appropriation to cover a temporary shortfall in another, even when reimbursement is contemplated.
- All items of expenditure are classified as either expenses or investments. An expense determination means an expense-type appropriation (like O&M) must be used; an investment requires a procurement or construction appropriation.
- One is prohibited from doing indirectly what an appropriation does not permit to be done directly.
- The *necessary expense test* is the principal legal test for the purpose characteristic (see Box 11.2).

One must be careful in applying this test to properly construe the meaning of the word *necessary*. The object of expense need only bear a logical relationship to the appropriation; it must contribute materially to the accomplishment of the agency's mission. One should not impute on the necessary expense rule a requirement for efficiency or economy. One is not limited to the most efficient or best alternative under the necessary expense rule. However, a logical connection cannot be construed too broadly. The tastes and desires of a senior official, a best practice from private industry, or the importance of the item do not, in and of themselves, constitute a logical connection.

The other two components of the necessary expense test are relatively straightforward. Objects of expense that are prohibited by other law, despite having a logical relationship to the agency's function and a given appropriation,

BOX 11.2 NECESSARY EXPENSE TEST

Under the necessary expense test, for an expenditure to be justified, it must 1) bear a logical relationship to the appropriation charged, 2) not be prohibited by law, and 3) not be otherwise provided for. That is, it does not fall within the scope of some other appropriation.

are not permitted to be funded by that appropriation. The purchase of a foreign-made component, regardless of how logically related to one's mission, if it violates a "Buy American" provision, is a violation of fiscal law. Lastly, one cannot charge an appropriation for an object of expense if a more specific appropriation exists for that expense. For example, the Navy was prohibited from paying for the dredging of a channel to support the construction of a ship from the ship's construction appropriation because dredging was specifically addressed in an Army Corps of Engineers appropriation. Regardless of whether the expense fit logically within the shipbuilding appropriation, this expense was "otherwise provided for" [11].

TIME. It is important to recall that an appropriation provides two authorities: to enter into obligations and to make payments, or expenditures, from the Treasury to pay those obligations. The general rule is that the availability relates to the authority to obligate the appropriation and does not necessarily prohibit payments after the expiration date for obligations previously incurred. Figure 11.2 defines key terms and authorities related to the time characteristic of appropriations. Of note is that the duration of the obligation period varies; some appropriations are annual (e.g., O&M); others are multiyear (e.g., procurement which normally has a three-year obligation period); yet others are no year [e.g., base realignment and closure (BRAC) funds and working capital funds]. The important legal test for whether one obligates funds in the proper period is the *bona fide needs rule* (Box 11.3). Bona fide need questions are all facets of the same basic concern—whether an obligation bears a sufficient relationship to the legitimate needs of the period of obligation availability of the appropriation charged.

Determination of what constitutes a bona fide need of a particular fiscal year depends on the facts and circumstances of the particular case. A common application of the rule is that an appropriation is not available for the needs of a future year. For example, suppose that as the end of a fiscal year approaches, an agency has remaining budget authority and purchases components for an item that is included in the subsequent year's procurement appropriation. This would be tempting for a program office that is concerned about keeping obligation rates on current funds high, and it would possibly buy some float in the production

BOX 11.3 BONA FIDE NEEDS RULE

The bona fide needs rule simply states that a fiscal-year appropriation can be obligated only to meet a legitimate, or bona fide, need arising in (or, in some cases, arising prior to but continuing to exist in) the fiscal year for which the appropriation was made.

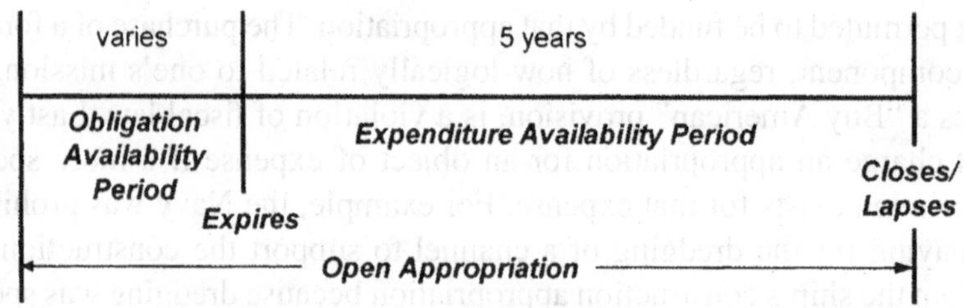

FY08 O&M Example: Opens on 1 Oct 2007
Expires on 30 Sep 2008
Closes on 30 Sep 2013.

Obligation Availability Period – the period during which new obligations may be incurred as specified by the appropriation. For annual appropriations, the period from 1 Oct through the following 30 Sep. At the end of the obligation availability period, the appropriation is said to ***Expire***. Expired authority is budget authority that is no longer available to incur new obligations, but is available during the five-year ***Expenditure Availability Period*** for disbursement of obligations properly incurred during the budget authority's period of availability or to cover legitimate obligation adjustments or for obligations properly incurred during the budget authority's period of availability that the agency failed to record. At the end of the five-year period the appropriation ***Closes*** or ***Lapses***. After the appropriation closes, all obligated and unobligated balances are canceled and all remaining funds are returned to the general fund of the Treasury and are thereafter no longer available for any purpose. Any legitimate invoice that is presented for payment after the appropriation closes must be paid from currently available funds.

Fig. 11.2 Appropriations terminology.

schedule by obtaining materials a little early. On the surface, it appears to be good program stewardship. Unfortunately, the purchase would violate the bona fide needs rule because the items are not a current-year need, but rather are a documented need of the subsequent year (evidenced by the budget).

Time questions are often related to the concept of severability. What is meant by *severable*? According to the Comptroller General, "The determining factor for whether services are severable or entire is whether they represent a single undertaking" ([12], pp. 5–24). Ongoing services, such as groundskeeping, engineering and logistics support, and leases are considered severable because the work done in one period can be separated, or severed, from the work done in another period, and the government still receives benefit for the work previously performed. The functional opposite of severable is *entire*. A task or item of expense is said to be entire if it represents a single undertaking: overhauling a pump, building a missile, constructing a building, or purchasing a computer are all entire tasks. A half-overhauled pump is of no value; a partially manufactured missile meets no need of the government. In other words, the government's needs are not met by a segment of the work.

This concept of severability vs entirety has significant financial implications. In brief, tasks that are severable are *incrementally funded* in annual installments; tasks that are entire are *fully funded* regardless of how long they take to accomplish. Incremental funding means that an effort is financed in periodic (normally annual) portions. The policy exists to conform to the bona fide need rule because funding a project beyond the end of the fiscal year (FY) creates a mismatch between the timing of the need and the fiscal year of the funds. It is also DoD policy that "annual budget estimates for Research, Development, Test and Evaluation (RDT&E) projects and programs, including developmental and operational test and evaluation programs, are to be prepared on an incrementally programmed basis" ([13], pp. 1–42). Whereas an end item, such as a prototype missile, would ordinarily be considered entire and be fully funded (see the following for discussion of full funding), because it is an R&D item, it would be incrementally budgeted and funded. Let us look at a hypothetical example (Box 11.4). Here we have a prototype missile being designed, built, and tested over a four-year period, with portions of the activity (and, hence, cost) being assigned to different periods. Under incremental funding, each year's cost would be budgeted and funded separately: $2.0 million in FY07, $4.5 million in FY08, $5.1 million in FY09, and $0.3 million in FY10.

Full funding, on the other hand, means that the entire cost to procure an item is budgeted, and those funds are obligated at the time the capability is ordered. Full funding conforms to the directive that "an end item budgeted in a fiscal year cannot depend upon a future year's funding to complete the procurement" ([13], pp. 1–24). Full funding requires that each year's appropriation request includes the funds estimated to cover the total cost of usable end items, such as a missile or tank. As an example, let us say we have an FY07

BOX 11.4 INCREMENTAL FUNDING EXAMPLE

Build and test a prototype missile	FY07	FY08	FY09	FY10	Total
Design	2.0	3.0	—	—	
Build	—	1.5	5.0	—	
Test	—	—	0.1	0.3	
Total costs	2.0	4.5	5.1	0.3	11.9
FYs of funding	2.0	4.5	5.1	0.3	11.9

BOX 11.5 FULL-FUNDING EXAMPLE

Procure 50 production missiles*	FY07	FY08	FY09	FY10	Total
Prime contractor	1.0	3.0	3.5	0.5	
Guidance section	0.5	2.0	2.3	0.6	
Govt. testing	—	—	0.5	0.2	
Total costs	1.5	5.0	6.3	1.3	14.1
FY of funding	14.1				

*Vehicles to be delivered December 2009–November 2010

requirement to build 50 missiles. Those missiles have a procurement lead time such that deliveries will begin in December 2009. Costs are shown in the Box 11.5 and spread over the four years. The full cost of $14.1 million will be budgeted in FY07 and obligated in FY07—even though the contractor will expend those funds over the next four to five years.

There are a few exceptions to full funding, including *advance procurement* and *multiyear procurement*. Advance procurement involves budgeting for long lead-time items in advance of the fiscal year in which the end item is fully budgeted. The funds are added to the budget authority for the prior fiscal year and deducted from the budget authority of the fiscal year for which the project is identified. In the example in Box 11.6, a submarine is budgeted in FY09 for a cost of $2.5 billion with $300 million in advance procurement. The program office will receive $300 million in FY08 to buy the long lead-time material. Their FY09 budget justifies the full funding of $2.5 billion for the submarine, but acknowledges the advance procurement and requests the balance of $2.2 billion.

Multiyear procurement (MYP) is a contractual commitment for support of outyear end items and is an exception to the bona fide need rule and full-funding policy, so it requires congressional approval. MYP is normally

BOX 11.6 ADVANCE-PROCUREMENT-FUNDING EXAMPLE

Build a submarine in FY09	FY08	FY09
FY08 adv proc for FY09	0.3	
FY09 SCN		2.5
Less adv proc		(0.3)
Total received	0.3	2.2

used when the production risk is minimal and opportunities exist for economies of scale with a contract for a large quantity. Annual increments for that procurement are still budgeted and obligated, but the government enjoys the lower unit cost, and the industry partner enjoys stability.

AMOUNT. In addition to specifying the purposes for which an appropriation can be cited and the period of time in which it is available, the amount available to an agency further constrains that agency's discretion. Restrictions on amount are not as simple as a prohibition against spending two million dollars when an agency has only appropriated one million. The restrictions are concerned more generally with spending money one does not (yet) have. The Antideficiency Act is the principal statute that addresses the amount characteristic.

The Antideficiency Act prohibits the following:

- Making or authorizing an obligation or expenditure from any appropriation (or apportionment) in excess of the amount available in the appropriation (or apportionment) {[10], § 1341(a)(1)(A) and § 1517(a)}
- Involving the government in any contract or other obligation for the payment of money for any purpose in advance of appropriations made for such purpose {[10], § 1341(a)(1)(B)}
- Accepting voluntary services for the United States, or employing personal services in excess of that authorized by law, except in cases of emergency involving the safety of human life or the protection of property {[10], § 1342; [11], pp. 6–36, 6–37}

The first is the commonly held definition of the Antideficiency Act—that one cannot spend more than one has. (An apportionment is a legally defined portion of an appropriation.) The second is also fairly common and most often occurs when appropriations are delayed as a result of continuing resolutions, the funds are being held by a comptroller for administrative purposes, or a contract is signed late in one fiscal year for work that is substantively a need of the subsequent fiscal year. The third item is not a prohibition against having the Boy Scouts paint the airplane at the entrance to the Air Base; that is a gift, not a voluntary service. This one prohibits the contractor from working "at risk" pending receipt of the appropriation or a signed contract. The difference is that the Boy Scouts do not expect to get paid; the contractor does. If the contractor works at risk and then bills later for that time, there has been a violation of the Antideficiency Act: the government is obliged to pay for services received in the absence of valid budget authority.

Sometimes the level of funding and the requirement are misaligned. Contingent events occur; risk is experienced; technical problems plague the program; or any of dozens of other events place the program in a position where the amount and type of funding is misaligned with the tasks that need to be performed or expenses that need to be borne. There is some flexibility granted in most appropriations to move a little money around. In short, a *transfer* is moving funds from one appropriation title to another; *reprogramming* is moving funds within an appropriation title from one subaccount to another. Shifting RDT&E, Navy to Aircraft Procurement, Navy is a transfer. Shifting Aircraft Procurement, Navy BA1 (buying tactical aircraft) funds to BA6 funds (buying initial outfitting spares) is a reprogramming. Each year, Congress gives the Secretary of Defense (SECDEF) general transfer authority for high-priority requirements that require the secretary to notify OMB and Congress. Congress reserves the right to disallow or modify a transfer if its members disagree with the amount or the intent.

11.4.2 MANAGEMENT FRAMEWORK

Fiscal law is inherently conservative; indeed, it can be said that the overarching virtue in the budgeting and comptroller world is consistency. If the requirement, the plan to meet it, and the pace of spending are consistent, then a program is considered financially healthy. It is when things change that scrutiny and risk increases. Unfortunately, it is the nature of acquisition that things are not consistent. Things change deliberately, and things change unexpectedly. Through the program changes, the program-management team members must be cognizant of how they expend funds.

One of the principal financial considerations of the program office is the obligation rate for budget authority. Each program creates a spend plan (also known as a financial plan or a phasing plan), and adherence to that plan is expected. Overobligation and underobligation are viewed as evidence of program problems. Overobligation could be interpreted as excessive cost growth or increased spending to deal with schedule slippage or technical problems; underobligation could be interpreted as a lack of progress or an overabundance of resources. The program either needs to spend the funding at the rate specified or have ready an acceptable explanation why it has not. This attention to the rate of spending will occasionally rush the program to contract award, perhaps prematurely.

Illustrative obligation and expenditure rates are provided in Table 11.2. Annual, expense-type appropriations such as O&M have very fast obligation and expenditure rates. Because they are one-year appropriations, one obviously expects them to be 100% obligated the first year. However, because of items with long lead times, the dynamics of end-of-year spending, and delayed invoicing, about half the obligated funds expend in the first year; the bulk of the balance expends in the second. Contrast this to investment-type appropriations like aircraft procurement and shipbuilding. Their obligation availability periods are three and five years, respectively, but one still expects the majority of funds to be obligated the first year. The logic in this is, "if you didn't need it this year, why did you budget for it?" Expenditure rates, though, tend to be quite slow given the time it takes to build a ship.

TABLE 11.2 OLIGATION AND EXPENDITURE RATES

Appropriation		CY, %	CY + 1, %	CY + 2, %	CY + 3, %	CY + 4, %	CY + 5, %
O&M	Ob	100	—	—	—	—	—
	Ex	75	94	97	99	100	—
MilPers	Ob	100	—	—	—	—	—
	Ex	98	100	100	—	—	—
RDT&E	Ob	93	100	—	—	—	—
	Ex	57	92	98	99	99	100
Aircraft	Ob	84	96	100	—	—	—
Procurement	Ex	25	63	88	95	98	99
Shipbuilding	Ob	63	82	90	96	100	—
	Ex	14	40	60	75	88	96

By maintaining a management reserve of funds as risk mitigation, a program manager impacts obligation rates. In the current policy climate, a program cannot explicitly incorporate management reserve in a budget. Yet, studies have demonstrated that there are benefits to maintaining a certain level of "slack" in one's budget. (*Slack* in a budget is similar to the concept of "float" in a schedule.) Slack permits a program to respond to contingencies, and it allows for additional planning and control activities without taking from production [14, 15]. In short, budgetary slack mitigates many forms of risk. So the management team must create slack. It can be done externally or internally. Externally, by DoD policy, the program might budget for engineering change orders at an amount commensurate with the risk in the program. If the program finds it is continually short of funds, then perhaps it is not adequately assessing risk and could budget more for change orders. Internally, a common method for creating slack is to conservatively estimate the cost of the activities in the budget, but to hold those conducting such activities to a more aggressive plan, that is, to budget for a most likely cost, but to allocate only 90% to the team with instructions to seek ways to more efficiently manage the effort. If the team is successful, the remaining 10% can be used by the program manager for other emergent needs. If the team is not successful, the manager still has the 10% balance to bail them out. Chances are that some program activities will create slack for other activities that were not so fortunate. Program managers must continuously assess financial risk and mitigate that risk by exploiting financial opportunities. Unfortunately, most programs understand that their chain of command is doing the same thing to them: holding back a percentage as risk mitigation, so that they already start a little behind!

The program office must remain vigilant to the state of future funding. This is not just attentiveness to the POM or budget; it is a year-long requirement. In the spring, the Congress is reviewing next year's budget and holding hearings, while the future years' POM building process is in full effect. In the summer, the Congress is enacting the next year's budget, while the service or component staff is reviewing the future years' POM and the budget submission. In the fall, the Congress is authorizing and appropriating the new fiscal year's funds while the future years' POM and budget are being reviewed at the OSD level. Finally, in the winter, the President's budget is submitted, and the cycle resumes. At any level, at almost any time, changes can occur that affect current execution and future needs.

Finally, the program office must simultaneously manage multiple years of financial information as illustrated in Figure 11.3. As stewards of the public's money, defense managers are called to account for past spending. There exists legal responsibility to manage appropriations for years after they have expired for obligational purposes. One must also actively manage the current year funds and articulate and justify future years' needs.

11.4.3 CONCLUSION

There are actors and processes related to financial management that simultaneously support the objectives of acquisition and support the objectives of long-standing notions of political control and accountability. Sometimes these objectives appear to conflict. Program progress and success are occasionally stymied by the limits placed on funding. Those limits are characterized by the purpose, time frame, and amount of funds appropriated.

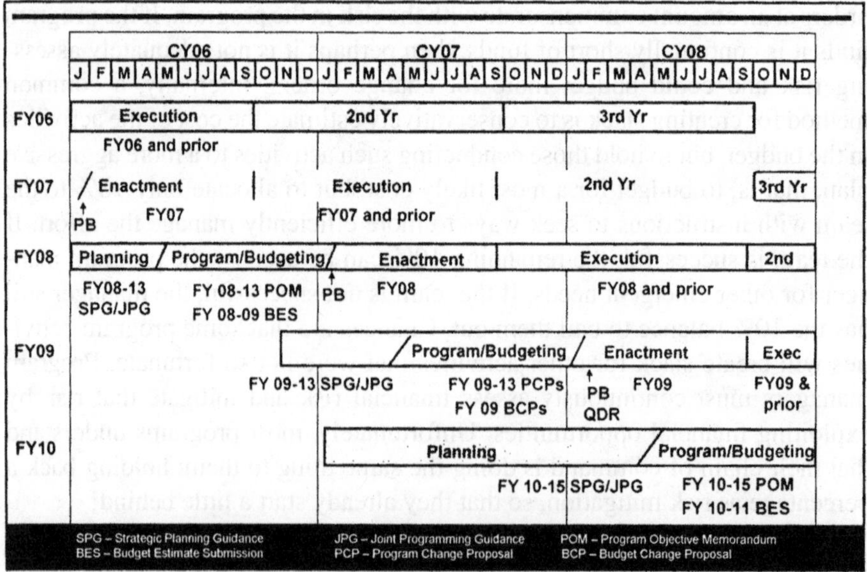

Fig. 11.3 Overlap of annual financial cycles.

Not only is it a goal of the DoD to acquire a system capable of supporting military operations, but it is equally a goal of the DoD to comply with the laws concerning how taxpayer dollars may be used. The goal of the various actors involved is to ensure both objectives are met simultaneously. It should not be the case that compliance interferes with program success, nor should programs violate fiscal law in pursuit of success.

11.5 LOOKING OUTSIDE THE DOD

The Department of Defense does not operate independently. Like all activities of the executive branch of the federal government, it is constrained and monitored by the legislature and is accountable to the public at large. With respect to defense acquisition, Congress is a vital external stakeholder. The United States is a nation of laws, and those laws not only constrain how defense acquisition is conducted, but authorize the acquisition program to exist in the first place. Congress further plays a key oversight role in the system of checks and balances. A successful acquisition manager understands both roles of the legislature and considers them in managing the program. Another vital external stakeholder is the defense industry, the for-profit partners with the DoD in the development, fielding, and support of systems. Although a discussion of the "iron triangle" could be (and has been) the subject of an entire book, this section of the chapter briefly examines how these two external stakeholders affect the financial management of major defense-acquisition programs.

11.5.1 CONGRESS

Yes, ours is a nation of laws, and that certainly holds true in the realm of defense acquisition. Programs are authorized, and the funds to pay for them are provided in law. Management practices affecting things like human resource policies, the proper use of funds, and contracting methods to deliver programs are prescribed by law. Legislation is the first of two important functions performed by the Congress. All legislation follows the same basic process as it winds through Congress. Most of the work of drafting or marking up legislation is done in committees. The committees often hold hearings about the proposed legislation, inviting testimony from experts from the executive branch, private industry, nonprofit special interest or watchdog groups, and congressional agencies such as the GAO, Congressional Research Service (CRS), or Congressional Budget Office (CBO). A bill is influenced by those invited to testify in hearings, by the committee which drafts and marks up the bill, by others outside the committee who offer amendments, and by the conference committee. At any of these stages, language affecting programs can be added, deleted, or modified. Figure 11.4, courtesy of the

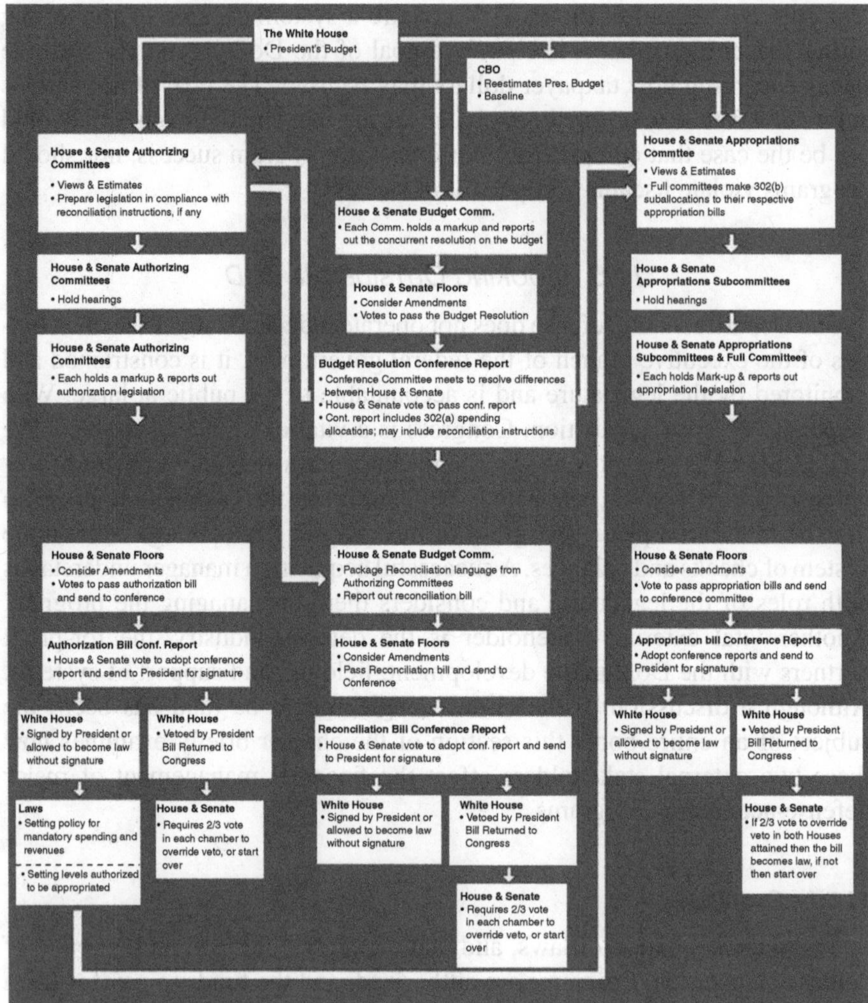

Fig. 11.4 The appropriations and budget process (http://budget.senate.gov/republican/analysis/budgetprocess.pdf).

Senate Budget Committee, displays the legislative process for the various products that affect defense budgeting and acquisition.

The two annual laws that most directly affect a project/program manager (PM) are the authorization act and the appropriations act. Generally, an authorization act "establishes and continues the operation of a federal program or agency either indefinitely or for a specific period or that sanctions a particular type of obligation or expenditure within a program" ([17], p. 15). The authorization act permits programs to exist, defines their scope (e.g., quantity), can set programmatic restrictions, and authorizes (but does not provide) appropriations at a certain level. The appropriation act actually provides the funding to carry out the authorized program.

Next are excerpts from the Fiscal Year 2007 Defense Authorization Act [18], which illustrate the authorization of appropriations as well as specific authorizations and constraints placed on programs:

SEC. 101. ARMY.

Funds are hereby authorized to be appropriated for fiscal year 2007 for procurement for the Army as follows:

(1) For aircraft, $3,451,429,000.
(2) For missiles, $1,328,859,000.
(3) For weapons and tracked combat vehicles, $2,278,604,000.
(4) For ammunition, $1,984,325,000.
(5) For other procurement, $7,687,502,000.
(6) For National Guard Equipment, $318,000,000.

SEC. 112. MULTIYEAR PROCUREMENT AUTHORITY FOR MH–60R HELICOPTERS AND MISSION EQUIPMENT.

(a) MH–60R HELICOPTER—Subject to subsection (c), the Secretary of the Army, acting as executive agent for the Department of the Navy, may enter into a multiyear contract for the procurement of MH–60R helicopters.
(b) MH–60R HELICOPTER MISSION EQUIPMENT—Subject to subsection (c), the Secretary of the Navy may enter into a multiyear contract for the procurement of MH–60R helicopter mission equipment for the helicopters covered by a multiyear contract under subsection (a).
(c) CONTRACT REQUIREMENTS—Any multiyear contract under this section—

(1) shall be entered into in accordance with section 2306b of title 10, United States Code, and shall commence with the fiscal year 2007 program year; and
(2) shall provide that any obligation of the United States to make a payment under the contract is subject to the availability of appropriations for that purpose.

Although this language is rather general, accompanying the authorization act is the report of the conference committee and a joint explanatory statement. Both documents—although they do not enjoy the force of law—amplify and clarify what Congress intended in the act itself. Generally, the DoD complies with these conference reports and statements to avoid raising the ire of Congress, but it is not required to.

In a perfect legislative world, the authorization act authorizes, and the appropriations act funds; yet, in reality, the authorization limits the scope and size of appropriations, and appropriations contain legislation beyond the provision of budget authority. In some cases, the appropriation bill provides amounts in excess of the authorization or less than the authorization. Generally, a program authorized in the appropriation might expend the money, but an appropriation that has restrictions in the authorization bill is bound by those restrictions [19]. There are times when there is apparent conflict between the two bills. In those cases, the PM should consult the acts' conference reports and explanatory statements as well as confer with the service's Office of Legislative Affairs or Office of General Counsel to ensure all programmatic requirements are met and all funds are spent in accordance with the law. In the event of conflict between legislation, affected funds might not be released to the program office until that conflict is resolved.

Program managers and acquisition officials often complain that Congress micromanages. The DoD managers believe, for example, they should decide on matters regarding the use of award and incentive fees, where raw materials can be purchased, precisely how many B-52s and attack submarines the DoD should keep, and whether or not to use contractors as lead system integrators—but all of those matters were addressed in recent legislation, restricting some of that inherent authority to manage. Why does Congress appear to micromanage, and through what means does it do it?

The second important function is oversight. Congress oversees, or, in extreme cases, micromanages, the affairs of the executive branch in a number of ways and for a number of reasons. First among those reasons is that Congress has the Constitutional power in Article I, Sec. 9, to "make rules for the government and regulation of the land and naval forces." When that power is wielded, it can appear to be micromanagement. Second, Congress might desire to shape defense policy in ways that differ from the executive branch, such as whether to build the executive transport helicopter overseas. The U.S. Congress can respond to media publicity of defense matters, especially in the case of apparent mismanagement. In the 1980s, it was apparent $600 toilet seats; in 2006, it was the condition of Walter Reed Medical Center. Micromanagement can result from partisan competition or congressional-executive branch competition, as well as mistrust. Because the defense budget has historically accounted for at least half of all federal discretionary spending, it is a favorite target for those legislators seeking to influence federal spending, even for nondefense matters. For instance, spending that might better fit in a Health and Human Services bill (e.g., breast cancer research) is sometimes added to the defense bill because a million-dollar earmark is less noticeable and less disruptive there. A final reason for congressional activism is the advocacy or protection of constituent interests, e.g., military installations, labor, or defense contractors [20].

It is not only important to examine why Congress tends to control defense, but it is useful to examine the manner in which such control occurs. Research provides us with a framework of tools, including restrictions on the use of funding, tools for gathering information, and accounting requirements [20]. For example, Congress might earmark funds for specific purposes, as in this example from the Fiscal Year 2007 Defense Appropriations Act [18]:

> SEC. 8022. (a) Of the funds made available in this Act, not less than $35,975,000 shall be available for the Civil Air Patrol Corporation, of which—
>
> (1) $25,087,000 shall be available from "Operation and Maintenance, Air Force" to support Civil Air Patrol Corporation operation and maintenance, readiness, counterdrug activities, and drug demand reduction activities involving youth programs;
>
> (2) $10,193,000 shall be available from "Aircraft Procurement, Air Force"; and
>
> (3) $695,000 shall be available from "Other Procurement, Air Force" for vehicle procurement.

Congress might place restrictions on the reprogramming and transfer of funds between accounts, or it might restrict funds pending executive compliance with provisions of law, as in this example:

> SEC. 8008. None of the funds provided in this Act shall be available to initiate:
>
> (1) a multiyear contract that employs economic order quantity procurement in excess of $20,000,000 in any 1 year of the contract or that includes an unfunded contingent liability in excess of $20,000,000; or
>
> (2) a contract for advance procurement leading to a multiyear contract that employs economic order quantity procurement in excess of $20,000,000 in any 1 year, unless the congressional defense committees have been notified at least 30 days in advance of the proposed contract award [18].

Congressional committees and individual members can gather information formally in hearings and informally outside of hearings or through the use of reviews, audits, and investigations by committee staffs, the GAO, and CRS. Each year, the GAO publishes over 100 reports about the DoD. It might also require the DoD to submit reports. Finally, Congress can place requirements on program execution:

> SEC. 8061. None of the funds appropriated or made available in this Act to the Department of the Navy shall be used to develop, lease or procure the T–AKE class of ships unless the main propulsion diesel engines and propulsors are manufactured in the United States by a domestically operated entity [18].

The effective program manager understands these two functions of Congress and manages as well as possible in light of them. Just as there is technical risk to every program, there is political and budgetary risk, as well.

11.5.2 INDUSTRY

We have a representative government in a nation that embraces democratic ideals. The Congress tends to reflect the will of the people and of organizations of people in our society. Sometimes Congress is proactive, generating a response from society; sometimes it is reactive to grass roots or organized effort. One segment of society that is particularly important concerning defense acquisition matters is industry. Individual firms, as well as trade associations and lobby groups, will attempt to influence the decision-making processes within government. The flip side of this, as those doing the business of the federal government, is that defense acquisition managers must rely on these firms to carry out that business—to do the engineering, program management, integration, construction, testing, and support of systems. Understanding how industry works is as important as understanding how Congress works.

Congress is not only responsible for raising and supporting armies and providing and maintaining a navy (Article I, § 8), but it must also concern itself with many other issues. Among those issues are labor laws, the macroeconomic condition of the country, the protection and development of strategically and economically important industries, health care, and other pressing matters. Individual firms and trade groups will pressure the legislature to pass laws that favor their interests. Those laws might include protections from foreign competition, for example. "Buy American" provisions are designed to protect the interests of U.S. suppliers, even though consumers might have to pay more for the those products. And sometimes that consumer is a DoD program manager.

Construction projects are particularly important to a member of Congress because they not only result in better facilities in that district (which could mean more jobs or more money), but the construction project itself employs local workers. This applies not only to the construction of buildings, but to the manufacture of aircraft, tanks, ships, missiles, and other capital goods. Yes, Congress is concerned that a well-balanced and effective military is equipped, but it is also concerned that industry is vibrant, jobs are plentiful, and the economy is strong—particularly among a congressman's particular constituency. Thus, decisions are often made that appear to be suboptimal to the DoD, but which might be optimal from a broader perspective.

Shipbuilding is one example of the preceding situation. The Navy produces a 30-year shipbuilding plan that shows the evolving size and structure of the naval force—an important document for Navy planners, for the shipbuilding industry as it plans, and for Congress as it formulates budgets and sets national policy. There is a shipbuilding association that testifies each year before the armed services committees, commenting on their capacity to meet the Navy's plan and, often, arguing for more business. There is also a shipbuilding caucus in Congress staffed by members whose districts and states include shipbuilding concerns. To most successfully budget and manage, defense program managers, PEOs, and acquisition executives should be cognizant of who is a member of these organizations, what positions they hold, and how they are likely to react to changes in the Navy's shipbuilding plans. When formulating plans, as well as executing them, the services and industry need each other to succeed, but their goals are often not aligned.

The program manager needs to invest time and attention to understanding his industry partners. That partner wants to produce good-quality products at a fair price, but has other concerns, too. Are those concerns profit margin, or return on assets, or return on investment, or cash flow? Are short-term earnings more important than market position? Is the contractor desperately underbidding to win work, but adding risk of cost growth? Does the contractor have the financial resources to sustain efforts until progress payments begin? By understanding that partner and what is important to him, the PM can shape the relationship through contractual and monitoring mechanisms to ensure both parties enjoy a successful partnership: the military gets a quality product at a fair price, and the contractor makes a reasonable amount of money.

In some respects, the DoD is Congress's agent. Congress sets policies, writes laws, monitors, and requires reports; Congress's goals and information needs differ from the DoD's, and both parties attempt to maximize their desired outcomes. Likewise, industry is the DoD's agent. Here, again, both parties attempt to maximize their outcomes at the lowest cost. The program office monitors contractor billing and performance, and the contractor attempts to contain costs and maximize revenue to increase profits. Interestingly, in our democratic society, industry looks at Congress as an agent—as one who is in a position to serve its interests. This iron triangle is a complex web of political, economic, and social relationships. Each actor must manage its pieces of it well to ensure that at the end of the day the military forces, shareholders, and constituents get a valuable outcome at a reasonable cost.

11.6 TIMELESS QUESTIONS IN DEFENSE FINANCIAL MANAGEMENT

Thus far, this chapter has highlighted the financial aspects of the management of major defense-acquisition programs. At this point, the perspicacious students might have formulated a number of questions:

1) Why are budgets not more stable and more aligned with the dynamics of the program rather than the dynamics of the annual political process? Specifically, why not separate the capital budget for acquisition from the

annual operating and support budget like most companies and many other governments do?

2) Why is there never enough money? What are the pressures on the defense top line that affect the amount available for development and procurement?

3) Why do we seem to perpetually reform the accounting and management information systems to improve these processes? Should they be fixed by now?

These are not new questions. Indeed, some have vexed the DoD and governments for decades. The premise of this textbook is that acquisition reform policies evolve in response to contemporary initiatives, but that certain concepts are constant and can be seen as anchors in an acquisition sea of change. Some reform policies over the years have attempted to address entrenched problems in different ways. This section of the chapter will address the entrenched issues, not to propose yet another reform, but to help the acquisition program managers who face them to better understand the underlying issues and why the problems are so persistent.

11.6.1 Problems with Budgeting

Nearly 70 years ago, V. O. Key wrote that the basic budgeting problem is: "On what basis shall it be decided to allocate *x* dollars to activity A instead of activity B?" [21]. Since then, no one has been able to adequately answer that question. A classical microeconomist does not see this as a problem and would respond with a discussion of marginal utility: if adding $1 more to activity A generates more utility than adding $1 to activity B, then add it to activity A until you have reached the equilibrium point. The problem is, no one actually thinks or behaves this way. Further, who shall define what a unit of utility is for activities A and B? Can they even be defined in comparable terms? Who is to say that military aircraft should get any money at all? Why not put the money into diplomatic efforts or economic ones instead? At the time this chapter is being written, Congress and the DoD are debating whether to invest in a second engine for the Joint Strike Fighter. It is not a simple economics question; it involves disparate views of risk and reward and international political considerations.

The PPBE process attempts to fund the best mix of programs that provide the right balance between near-term and long-term capabilities. Problems arise when there is more than one way to provide a capability, and each way involves different services or different factions within a service. Questions of fairness arise. Extant capabilities must continue to be funded or dismantled, further constraining the question. Other complications come from outside influences. If Congress dictates—like it did in the Fiscal Year 2007 Defense Appropriations Act—that the Air Force shall maintain a fleet of 93 B-52s, or the Navy will maintain 48 attack submarines, then those are no longer variables in the budget-allocation question, but rather constraints. If industry can only produce 12 ships a year, it does not matter if the Navy desires 14. Debates also exist in the balance between near-term and long-term readiness.

Budgets, therefore, tend to be incremental adjustments to the prior year's budget. It is the least controversial process and the one most likely to succeed. Thus, what is truly discretionary—the amount "in play"—in any given budget is a rather small slice [22]. The staunchest advocate of an evolutionary organizational theory would argue for it to be this way, but the evidence shows that budgets do *not* actually evolve smoothly. There are periods when big shifts occur. For instance, mine-resistant ambush protected (MRAP) vehicles are designed to counter the threat of improvised explosive devices in Iraq. In February 2007, the administration requested only a few million in the wartime supplemental for the vehicles, and the Army put $2.2 billion for MRAP on its FY2008 unfunded priorities list submitted to Congress. This request indicated a desire for the vehicles, but implied they were not a priority. Within the subsequent five months, however, the DoD named MRAP its highest priority program, reprogrammed over $1 billion of current year money into it, and amended the FY08 budget to include billions of dollars for approximately 20,000 vehicles. Was this an incremental adjustment to an evolving threat? Hardly. The threat from improvised explosive devices did not get 10 or 20 times worse in those five months. Clearly, budgets and resource allocation are not always incrementally adjusted. Sometimes big shocks occur to the system.

Private industry, as well as many state, municipal, and local governments, use one process to differentiate budget for capital items and another for operating expenses. Most people in managing their household budgets do the same; we apply one decision-making logic when budgeting for the electric bill or groceries and a different logic when deciding to purchase a new car or major appliance. Within the DoD, procurement budgets are generated alongside operating and salary budgets, and both are appropriated on an annual, line-item basis. Reformers have often questioned this practice. Most recently, the 2006 *Quadrennial Defense Review* (http://www.defenselink.mil/qdr/report/Report20060203.pdf) and the "Defense Acquisition Performance Assessment" [2] recommended a form of capital budgeting. Let us examine that debate.

Capital budgeting involves the analysis of costs and cash flows associated with an investment project. This analysis precedes and informs the decision to invest. At most levels of government (excepting, notably, the federal government) and in private industry, capital investment decisions are made in tandem with the decision of how to finance them; most public projects are paid for with special tax assessments, bond issuances, loans, or other forms of debt; private projects are funded with stock sales, bonds, venture capital, or by borrowing. There is a deliberate link between the investment and its implications on current and long-term budgets, fiscal policy, asset management, and cash flows [23]. Capital budgeting practices include life-cycle cost and benefit analyses. The costs include the obvious investment in the item, financing costs, and any incremental operating and support costs. Benefits include new revenue streams or lower operating and support costs. Investment decisions are based on one or more analytical techniques such as payback period, net present value, or internal rate of return. The capital budget and decision can be prepared outside the annual O&S budget cycle, instead of timing them according to project schedules.

Defense-acquisition reformers argue that such practices would provide much-needed stability to the management of defense-acquisition programs [24, 25]. Such stability, it is argued, will increase the likelihood of program success by addressing a significant extant risk factor. The "Fiscal Year 2007 Defense Acquisition Transformation Report" to Congress makes this argument ([26], p. 20). But the budgeting practices for capital investments at the federal level differ significantly from local government of industry. Capital budgets do not need to be separate from operating budgets in the absence of balanced budget requirements; the sovereign federal government can run deficits, raise revenues, print money, and borrow more readily than any other organization. Also, irrespective of the budgetary concern, federal capital projects are sometimes viewed as mechanisms for affecting macroeconomic conditions (e.g., interstate highway construction) and are essentially short-term goals achieved through a capital program, which confounds the analysis.

Such reform is problematic in the DoD for three main reasons. First, there is no benefit that is easily defined in financial terms. Computing a net present value or payback period is meaningless if the benefit and the cost cannot be expressed in consistent units. The DoD does not buy tanks because they generate cash flow. Yes, a cost-effectiveness analysis can be done for competing proposals that perform the same task, but when preparing capital budgets for items that generate disparate benefits, comparison is very problematic. Second, capital budgeting restricts policy choices by removing the decision to fund the capital investment from the larger discussion of overall defense policy and resource decisions. The legislature prefers to debate and decide defense matters once in the annual authorization and appropriation process. Separating the capital items can cause the same issues to be debated twice or can mask interrelationships that should be considered as a whole. Third, if done properly, PPBE should already consider life-cycle cost and independent cost analyses; and, if full funding is appropriated, then the goals of a capital budgeting process have already been met. Recent proposals to "fence" procurement funds in separate accounts do not provide the benefit of capital budgeting; in fact, they merely serve to restrict flexibility in execution.

What does this mean for acquisition managers? Although adopting capital budgeting reforms would potentially stabilize the funding outlook for select acquisition programs, such reforms are unlikely because they are politically untenable, add few new insights, and might actually inject additional rigidity into an already cumbersome process. Acquisition program managers should be aware of the debate, but should not view such reforms as a panacea or an external remedy to their funding stability problem. In a recent look at PPBE reform, Jones and McCaffery conclude, "those who plan, manage, and budget so as to allow legislators to help supply inputs to their vision are more likely to finish ahead in the battle of the budget" ([28], p. 19). Acquisition managers should be prepared to continue to expend time and attention on matters of acquiring and retaining financial resources.

11.6.2 TOP-LINE PRESSURES

Resource-allocation decisions are a question of who gets how much of the defense top line. Given that the cost of desired programs (or requirements, if you will) exceeds the available resources, this is a dilemma; indeed, this is problematic any time the top line is declining, flat, or rising below the pace of inflation. When there is such downward pressure on the top line, competition for resources is fiercer. What factors affect that top line?

Defense spending can be viewed in both absolute and relative terms. Figure 11.5 displays three measures of the defense top line over the period 1940–2011. In absolute terms, since WWII, defense outlays appear as a cyclical pattern—with a lower limit of about $350 billion and an upper limit of about $550 billion in FY2006 dollars. The cycles are fairly consistent and repeat every 15 to 21 years. Note that Fiscal Year 2007 is 18 years since the last peak in 1989, suggesting that *if the long-term pattern repeats*, defense can soon expect declining absolute top lines, which will continue for seven to ten years. In relative terms, since WWII, defense spending as a percentage of gross domestic product (GDP) has fallen rather steadily from a peak of 14.2% in 1952 to a low of 3% in 1999–2000. In 2006, defense spending stood at about 4.1% of GDP. Similarly, defense spending as a percentage of federal outlays has also fallen from a peak of 69.5% in 1954 to a low of 16.1% in 1999. *Should the long-term trends continue*, defense can expect declining relative top lines. But simply identifying the trend of the last 50 years does not mean that trend will apply to the next three to five years. More than a long-term trend is needed; there are external and internal factors that can affect the top line, apart from the obvious effects of the war.

Externally, defense spending is being "crowded out" of the federal budget as a portion of federal spending. In the 1950s, defense accounted for two-thirds of federal spending; by 2006, it fell to 20% of federal outlays. In contrast, Social Security and health care spending increased from about 15% of federal spending to over 40% in the same time period. The CBO projected that in the period from 2008–2017 defense budget authority will increase at

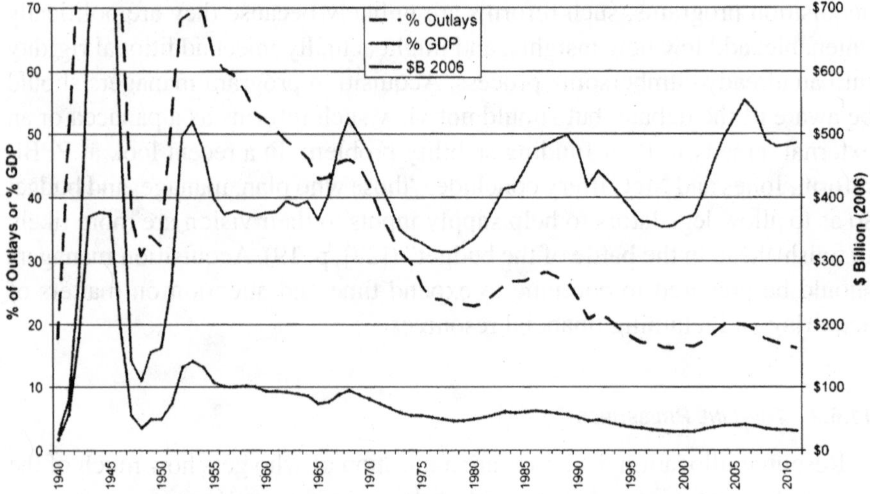

Fig. 11.5 Defense top line 1940–2011: Various measures [28].

2.2% per annum, relative to GDP growth of 4.5%, whereas mandatory spending will grow at 5.9% [29]

Defense top lines not only feel pressure from growing entitlement programs, but face pressure from other discretionary programs. From 1985 to 2006, total discretionary outlays as a percentage of GDP fell from 10.0 to 7.8%. Of those 2.2 percentage points, defense spending accounted for 2.1, and other discretionary spending accounted for 0.1 [29]. Nearly the entire reduction in discretionary spending was absorbed by defense. The last few years of that history show a modest recovery.

As of 2006, the United States had experienced four consecutive years of budget deficits following four consecutive years of budget surpluses. The Bush Administration's position in 2007 was that it will eliminate such annual deficits by 2012. When an administration or Congress wishes to reduce a deficit, generally it requires a combination of increasing revenues and decreasing outlays. If defense maintains even a steady proportion of federal spending during an overall decline, it will lose top line.

In addition to fiscal pressures, the defense top line responds to political pressures. Evidence strongly suggests that if the general public believes defense spending is too high, defense spending declines and vice versa. Studies conducted at the end of the Cold War demonstrated empirically that the desires of the public are reflected in future spending decisions [30, 31]. Figure 11.6 displays the direction and strength of public opinion about defense spending with subsequent changes in that spending. Change in defense spending correlates strongly with the direction and strength of public opinion. When public opinion favors increased defense spending, spending has tended to go up the following year; when the public favors a decrease, spending tends to drop the following year. The intensity of public opinion also forecasts the significance of the gain or drop. In combination with the data in Fig. 11.6, one might conclude that the public has a "comfort zone" of appropriate defense spending that ranges between $350 billion and $550 billion (FY2006 dollars). Yet, defense spending for FY08 is projected to be over $620 billion.

We see that, externally, there has been pressure on the defense top line from growth in mandatory spending accounts and nondefense discretionary accounts. The United States is again at a time when deficits are a point of concern, and as the war in Iraq progresses, public support of defense spending is waning. Time will tell how that manifests in spending decisions. External factors are vital when considering current defense management reform, but internal factors can be even more critical.

Internal factors affecting defense spending include the quality of budgeting and containment of growth in acquisition, personnel, and operation and maintenance expenses. Consider these data: from the low point of the 1990s' "procurement holiday" to 2006, the Navy's budget increased 46% in real terms, but the size of the fleet fell from 354 battleforce ships to 280; aircraft inventory fell from 2,559 to 2,330; and the number of personnel (uniformed and civilian) fell from 929,358 to 829,531 [33]. Whereas spending increased by half, the naval forces are 15% smaller. The last time defense experienced 12 continuous years of budget growth was 1979–1990, during which time the

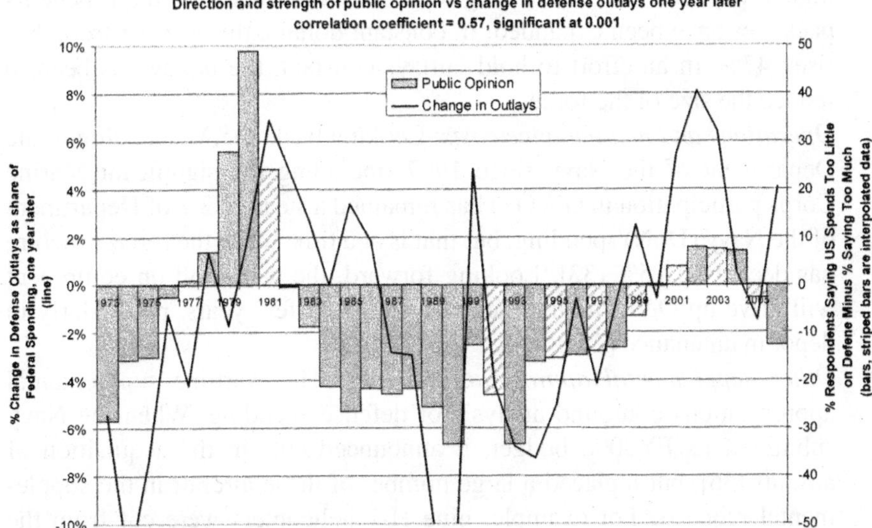

Fig. 11.6 Public opinion and defense spending, 1973–2006 (derived from [28 and 32]).

fleet grew from 530 to 587 ships. To sustain its current goal of a 313-ship Navy, there needs to be sufficient shipbuilding budget authority to consistently build an average of 11 ships a year. Since 1998, Navy five-year budgets have planned to build eight ships per year, but have succeeded in building only six ships per year. To get to 12 will require nearly doubling the annual shipbuilding budget, currently at $11 billion [34].

Some specific issues putting pressure on the defense top line from within the department include the following:

- *Acquisition costs*: The Department's appetite for major acquisition programs and the cost performance of those programs continue to be important issues. The Joint Strike Fighter, F-22A, the Army Future Combat System, the Air Force's Transformational Satellite System, the Navy's DD-1000 and Virginia-class submarine programs have all experienced significant cost growth, quantity reductions, or schedule slips [35]. In early 2007, the Navy acquisition executive issued a stop-work order for the relatively affordable Littoral Combat ship when the first ship in the class experienced costs 50% over budget. Thus, one comes to the difficult realization that just as other internal and external factors are putting pressure on the fiscal resources available for recapitalization, acquisition program performance itself is a source of fiscal stress.
- *Personnel costs*: Pay and benefits for personnel have increased in recent years. At the same time, accrual accounting changes have illuminated some costs (such as the accrued costs of retiree heath care) that have always been there, but were not explicitly recognized. The activation of tens of thousands of reservists and guardsmen for the Iraq war resulted in higher pay and increased long-term liabilities because their benefits packages have been expanded. In constant dollars, the cost per troop has risen 47%. In an effort to hold outlays constant, the policy has been to reduce the size of the force.
- *Operations and maintenance costs*: Looking back, O&M spending in the Department of the Navy since 1997 (including the significant Marine Corps participation in GWOT) has remained a steady 32% of Department of the Navy (DoN) spending, but that is at a time when the force structure has decreased 15% [33]. Looking forward, the war's toll on equipment will drive up O&M requirements for the next few years, particularly in depot maintenance [2].
- *Other supplemental appropriation issues*: The wartime supplemental appropriations confound analysis of defense spending. When the Navy submitted its FY2008 budget, it announced cuts in the acquisition of aircraft [36], but replaced a large number of those aircraft in the supplemental requests. For example, nine H-1 helicopters were cut from the base, but requested in the supplemental; eight H-60 helicopters were cut

from the base, but nine were requested in the supplemental; four V-22 tilt-rotor aircraft were cut from the base, and three were requested in the supplemental (despite the fact the V-22 had not yet been used in the war) [33]. Theses practices cause distortions in baseline budgets that can negatively affect future DoD budgets after the use of supplemental appropriations ceases, unless some type of "recapture" takes place.

Will the defense top line rise or fall over the next few years? Long-term historical trends suggest defense spending is likely to fall. External factors such as growing entitlement and discretionary programs restrict the room for growth in the defense budget, particularly when there is political pressure to reduce deficits. Internal factors push for a higher top line, but include significant inflationary effects; thus, top-line increases have bought less force structure (but not necessarily less capability). Supplemental appropriations are not a long-term source of relief and can even confound problems.

What does this mean for defense-acquisition managers? It appears the defense budget is under increased stress, as well as increased scrutiny, from both internal and external forces. If so, competition for resources for acquisition programs will become even more intense.

11.6.3 FINANCIAL-MANAGEMENT REFORM

While this text focuses on the constants in a sea of change, one of those constants *is* the sea of change. The effects of the currents in that sea should be a consideration of a program manager steering a course. In addition to episodic changes to acquisition policy, there are changes to financial management and budgeting and accounting policies and systems. Since the passage of the Chief Financial Officer's Act of 1990, Public Law 101–576, the DoD has struggled to produce accounting information that can withstand the scrutiny of an independent audit. Since 1995, DoD financial management has been on the Government Accountability Office's "high risk" list of those programs and operations most vulnerable to fraud, waste, abuse, and mismanagement [37].

The long-standing issues in defense financial management are primarily in the accounting realm; the DoD struggles to provide a clear and auditable accounting for the funds received. This is not surprising because the DoD's accounting and financial-management systems and those systems that feed them with transactional data were never originally designed to do this. The production of corporate-style auditable financial statements is a relatively new demand. Historically, such systems and their corresponding management attention were devoted solely to the process of building a budget, acquiring budget authority (appropriations), and accounting for that budget authority. They were financial systems built to ensure compliance with legal aspects of appropriations; they were not designed to provide management information beyond that narrow scope [38].

A goal of the reform agenda has been to develop systems that not only ensure defensible budgets and accountability for appropriations, but that also provide reliable, relevant, and timely cost and financial-management information for decision making. What it will take to affect that reform is a combination of changes to culture, incentives, and decision-making processes, which will require changes to information systems. Many reform efforts have started with information systems—technology—without changing the organization and have failed as a result. Recent plans seem to acknowledge those factors more explicitly than prior plans and might see greater success if they can overcome greater organizational resistance.

What does this mean for acquisition program managers? They will experience an environment of shifting currents in this continuously turbulent sea of change. Accounting and budgeting information technology systems will continue to evolve, which will distract the program office staffs. Until such time as the DoD is "managed in an efficient, business-like manner in which accurate, reliable, and timely financial information, affirmed by clean audit opinions, is available on a routine basis to support informed decision-making at all levels" [39], program managers should expect to be disrupted by, and to participate in, efforts to get there.

11.7 CONCLUSION

The intent of this chapter was two-fold. First, it was designed to give defense acquisition managers an appreciation for the reasons behind the complexity and sometimes apparent incoherence of the world of budgeting, cost analysis, appropriations, accounting, and financial management. It is hoped that now the reader will find this world less bewildering and more coherent. Hopefully, the student discovered there is a purpose behind the rules; that allocation decisions are not all that irrational; and that what makes for good business and good politics are not normally the same things. Second, this chapter is designed to show acquisition managers ways in which to use that information for the advancement of their programmatic goals. By way of analogy, recognizing that there is some structure and reason behind the complexity and somewhat incoherence of the game of chess is the first step toward becoming a proficient player. Knowing the rules of the game is vitally important if one is to win the game. To some degree, budgeting and financial management in defense can be viewed as a complex game.

This chapter outlined the basic financial management framework that every program office must work within. The basics of the processes used within the DoD for resource allocation (PPBE) were discussed, with an emphasis on those parts most critical to the program office: programming and budgeting.

We saw that obtaining and executing budget authority are closely linked. One must request budget authority in the amounts, in the form, and at the time that will legally permit the program office to execute that authority effectively. Program managers need to understand the key processes for authorizing programs and for appropriating funds to them, and they also need to understand the motivations behind, and mechanisms for, congressional oversight. Similarly, program managers should understand the motivations of their corporate partners. Finally, this chapter highlights the fact that the successful program manager must be aware of those factors in the environment that could affect program progress.

As much as program managers focus attention on and even bemoan financial management matters, the astute ones will recognize the importance. No program exists without financial resources. There are aspects of financial management in which becoming proficient at the game gives one an advantage. The successful manager will also do well to recognize those other aspects of financial management that are not problems to be solved, but rather constraints to be considered. Such an approach will decrease pressure, focus attention, improve risk management, and eventually lead to better outcomes.

REFERENCES

[1] Government Accountability Office, "Best Practices: Better Support of Weapon System Program Managers Needed to Improve Outcomes," Rept. to the Subcommittee on Readiness and Management Support, Committee on Armed Services, Rept. GAO-06-110, U.S. Senate, Washington, D.C., Nov. 2005.

[2] Defense Acquisition Performance Assessment Project Rept., Jan. 2006, p. 32; also, Government Accountability Office, "Defense Acquisitions: DoD has Paid Billions in Award and Incentive Fees Regardless of Acquisition Outcomes," Rept. GAO-06-66, Washington, D.C., 2006.

[3] Congressional Research Service, "Defense Acquisition: Overview, Issues, and Options for Congress," CRS Rept. for Congress, RL34026, Washington, D.C., June 2007.

[4] Defense Acquisition Transformation Rept. to Congress, Feb. 2007, p. 20; also, Government Accountability Office, "Defense Acquisitions: DOD Needs to Exert Management and Oversight to Better Control Acquisition of Services," Rept. GAO-07-359T, Washington, D.C., 2007.

[5] National Academy of Public Administration (NAPA), "Moving from Scorekeeper to Strategic Partner: Improving Financial Management in the Federal Government," Rept. for the U.S. House of Representatives, Committee on Government Reform, Subcommittee on Government Management, Finance and Accountability, Oct. 2006, Washington, D.C., 2006.

[6] Dept. of Defense, "What is a Budget?," Office of the Secretary of Defense Comptroller iCenter, Washington, D.C., 2006, http://www.defenselink.mil/comptroller/icenter/budget/whatisbudg.htm [retrieved 18 Dec. 2006].

[7] Dept. of the Navy, *Navy Budget Guidance Manual*, Part 1, p. I-2, http://www.finance.hq.navy.mil/fmc/PDF/partIR13.pdf [retrieved 23 May 2008].

[8] Dept. of Defense, *Financial Management Regulation*, Vol. 2B, Chap. 5, Sec. 050201, Washington, D.C., June 2006.

[9] Government Accountability Office, "Defense Acquisitions: Realistic Business Cases Needed to Execute Navy Shipbuilding Programs," Rept. GAO-07-943T, Washington, D.C., July 2007.
[10] United States Code, Title 31, U.S. Government Printing Office, Washington, D.C., http://www.access.gpo.gov/uscode/title31.html [retrieved 23 May 2008].
[11] Government Accounting Office, "Comptroller General Appropriations Decision," GAO B-139510, Washington, D.C., 13 May, 1959.
[12] Government Accountability Office, *Principles of Federal Appropriations Law*, 3rd ed., Washington, D.C., 2004, pp. 5–24.
[13] Dept. of Defense, *Financial Management Regulations*, Washington, D.C., 2007, http://www.defenselink.mil/comptroller/fmr/ [retrieved 23 April 2007].
[14] Glassberg, A., "Organizational Responses to Municipal Budget Decreases," *Public Administration Review*, Vol. 38, No. 4, 1978, pp. 325–332.
[15] Dunk, A., and Nouri, H., "Antecedents of Budgetary Slack: A Literature Review and Synthesis," *Journal of Accounting Literature*, Vol. 17, 1998, pp. 72–96.
[16] Senate Budget Committee, Washington, D.C.
[17] Government Accountability Office, "A Glossary of Terms Used in the Federal Budget Process," Rept. GAO-05-734SP, Washington, D.C., Sept. 2005.
[18] Fiscal Year 2007 Defense Authorization Act, Public Law 109–289, Washington, D.C., 2007.
[19] Congressional Research Service, "A Defense Budget Primer," Rept. RL30002, Washington, D.C., 1998, http://www.globalsecurity.org/military/library/report/crs/RL30002.pdf [retrieved 23 July 2007].
[20] Candreva, P. J., and Jones, L. R., "Congressional Control over Defense and Delegation of Authority in the Case of the Defense Emergency Response Fund," *Armed Forces & Society*, Vol. 32, No. 1, 2005, pp. 105–122.
[21] Key, V. O., "The Lack of a Budgetary Theory," *The American Political Science Review*, Vol. 34, No. 6, 1940, pp. 1137–1144.
[22] Wildavsky, A., and Caiden, N., *The New Politics of the Budgetary Process*, 4th ed., Addison Wesley Longman, New York, 2001.
[23] Lee, R. D., and Johnson, R. W., *Public Budgeting Systems*, 6th ed., Aspen Publishers, Gaithersburg, MD, 1998, pp. 363–378.
[24] Kadish, R., Abbott, G., Cappuccio, F., Hawley, R., Kern, P., and Kozlowski, D., "Defense Acquisition Performance Assessment Report for the Deputy Secretary of Defense," 2006, http://www.acq.osd.mil/dapaproject/ [retrieved 18 April 2007].
[25] McCaffery, J. L., and Jones, L. R., "Reform of Budgeting for Acquisition: Lessons from Private Sector Capital Budgeting for the Department of Defense," Rept. NPS-FM-06-029, Acquisition Research Sponsored Report Series, Naval Postgraduate School, Monterey, CA, 2006, http://acquisitionresearch.org/index.php?option=com_content&task=view&id=104&Itemid=41 [retrieved May 2007].
[26] Kreig, K. "Defense Acquisition Transformation Report to Congress," Washington, D.C., Feb. 2007.
[27] Jones, L. R., and McCaffery, J. L., "Reform of the Planning, Programming Budgeting System, and Management Control in the U.S. Department of Defense: Insights from Budget Theory," *Public Budgeting and Finance*, Vol. 25, No. 3, 2005, pp. 1–19.
[28] Office of Management and Budget, *Historical Tables, Budget of the United States Government for Fiscal Year 2008*, Government Printing Office, Washington, D.C., 2007.
[29] Congressional Budget Office, *The Budget and Economic Outlook: Fiscal Years 2008–2017*, Government Printing Office, Washington, D.C., Jan. 2007, p. 51.

[30] Hartley, T., and Russett, B., "Public Opinion and the Common Defense: Who Governs Military Spending in the United States?," *The American Political Science Review*, Vol. 86, No. 4, 1992, pp. 905–915.

[31] Higgs, R., and Kilduff, A., "Public Opinion: A Powerful Predictor of U.S. Defense Spending," *Defence Economics*, Vol. 4, No. 3, 1993, pp. 227–238.

[32] Smith, T. W., *Trends in National Spending Priorities, 1973–2006*, National Opinion Research Center, Chicago, 2007, p. 15.

[33] Dept. of the Navy, *Budget Highlights*, editions for Fiscal Years 1998–2008, Washington, D.C., 2007, http://www.finance.hq.navy.mil/fmb/08pres/books.htm [retrieved 13 Feb. 2007].

[34] Congressional Research Service, "Navy Force Structure and Shipbuilding Plans: Background and Issues for Congress," Rept. RL32665, Washington, D.C., Aug. 2006.

[35] Congressional Research Service, "Defense Budget: Long-term Challenges for FY2006 and Beyond," Rept. RL32877, Washington, D.C., April 2005.

[36] Castelli, C. J., "Navy Cuts Aviation Accounts, Budgets for Seven Ships in 2008," InsideDefense.com [retrieved 5 Feb. 2007].

[37] Government Accountability Office, "High Risk Series: An Update," Rept. GAO-05-207, Washington, D.C., Jan. 2005.

[38] Candreva, P. J., "Accounting for Transformation," *The Armed Forces Comptroller*, Vol. 49, No. 4, 2004, pp. 7–13.

[39] Rumsfeld, Donald H., "Financial Management Information in the Department of Defense," internal DoD Memorandum, Dept. of Defense, Washinton, D.C., 19 July, 2001.

STUDY QUESTIONS

11.1 Describe how and with whom the financial-management framework overlaps in the acquisition framework.

11.2 What are the characteristics of a "good" budget for an acquisition program?

11.3 It was said in this chapter that in the comptroller's view of the world, consistency is a virtue.

(a) In what ways is consistency beneficial?

(b) Should consistency be a virtue in acquisition program financial management, or is flexibility the greater virtue? Why?

11.4 What are some of the nonprogrammatic considerations that affect how funds are acquired and how they can be used by a program office?

11.5 How is the necessary expense rule a consideration as an acquisition program advances from demonstration to production?

11.6 How does the bona fide need rule affect budgeting and production scheduling for an acquisition program?

11.7 In what ways are the political decisions made by Congress in the authorization and appropriations processes a problem or a constraint for acquisition program managers?

11.8 What are the most significant financial-management issues a program must consider in its relationships with its industry partners during budgeting? During execution?

11.9 Describe the differences among cost, budget authority, and spending (expenditures)?

11.10 How is program management affected by strict adherence to obligation rate targets set by the component headquarters? What (dis)incentives does that create? How might waste occur?

11.11 Given the timeless questions in defense acquisition financial management, how should acquisition executives approach these problems? How should the individual program manager approach reform movements?

11.12 Consider your acquisition program in light of this chapter.

(a) What new insights do you have?

(b) What do you better appreciate or understand?

(c) What can you do to manage more effectively?

(d) How can your financial managers and cost analysts serve the program better?

(e) Which stakeholder relationships could be strengthened to better improve your financial situation?

Chapter 12

EARNED VALUE MANAGEMENT

KEITH F. SNIDER* AND JOHN T. DILLARD†
NAVAL POSTGRADUATE SCHOOL, MONTEREY, CALIFORNIA

LEARNING OBJECTIVES

- Understand the importance of *earned value management* as a project-management tool
- Describe how the *performance measurement baseline* is developed
- Given data for a simple contract, calculate *variances* and other *metrics* to evaluate contract performance to date and to forecast future performance

12.1 *INTRODUCTION*

This chapter discusses earned value management (EVM), an important tool for integrating cost, schedule, and performance measurements. Although EVM is widely used in project management in both the public and private sectors, its application in defense acquisition management is mainly in measuring performance on individual contracts. EVM provides "early warning" of contract cost, schedule, and performance issues, and it also predicts future contract performance based on past performance. Specifically, EVM provides for the following:

- Planning all work for the contract period of performance (POP)
- Integrating all contract work scope, cost, and schedule objectives in a single performance baseline
- Objectively assessing progress against that baseline
- Analyzing variances and forecasting their impacts
- Making meaningful information available to decision makers

---

*Associate Professor, Graduate School of Business and Public Policy.
†Senior Lecturer, Graduate School of Business and Public Policy.

12.2 EVM APPLICATION

EVM can be used across any type of commodity, but it is typically employed for "knowledge work" (such as software development or hardware design), in which the project results are less tangible than under lower-risk, production-type efforts. In such complex and risky projects, payments are usually made for work performed along the way, even though visible progress might be hard to gauge. Thus, EVM is most appropriate for contracts in which the government has a significant cost risk, that is, for cost-reimbursable rather than fixed-price or level-of-effort contracts.

The government includes in contracts the requirement for the contractor to report EVM data periodically (usually monthly). These data are required in significant detail for larger contracts and in less detail for smaller contracts.

BOX 12.1 EXAMPLES OF THE IMPORTANCE OF EARNED VALUE MANAGEMENT SYSTEMS

The Air Force Global Positioning System (GPS) Block II modernization program office had relied on earned-value-management reports to monitor the contractor's production efforts, but discovered in 2006 "that the contractor's earned value management reporting system was not accurately reporting cost and schedule performance data" ([1], p. 86).

"In March 2006, the lead contractor [for the Army's Armed Reconnaissance Helicopter program] lost its earned value management certification due to a recent compliance review that found lack of progress in addressing long-standing systemic deficiencies. Without certified earned value management data, the Army will not have timely information on the contractor's ability to perform work within estimated cost and schedule" ([1], p. 38).

"The restructuring of the Air Force Space-Based Infrared System (SBIRS) program modified the program's use of DOD's Earned Value Management System (EVMS). Specifically, Lockheed Martin and its subcontractors standardized EVMS procedures in an effort to provide more accurate and up-to-date reporting on the status of the program. In addition, an EVMS oversight team was established to focus on process improvements, and Lockheed Martin and its subcontractors developed a surveillance plan to review the EVMS data. The contractor is now monitoring EVMS data more closely through monthly meetings and reviews of specific cost accounts. Changes to the reporting of EMVS data also help identify risks more effectively" [2].

Because EVM uses data developed and reported by the contractor, its utility depends on the reliability and fidelity of that data (see Box 12.1). Consequently, a contractor's internal management information system will be expected to meet certain industry standards (currently, ANSI/EIA-748) to ensure it is capable of capturing and reporting quality data.

12.3 PERFORMANCE MEASUREMENT BASELINE

EVM measures actual contract performance (earned value, discussed next) against an integrated baseline, called the *performance measurement baseline* (PMB), which reflects a plan of cost, schedule, and technical progress for the POP. Obviously, the quality of those performance measurements is dependent in large part on the quality of the baseline itself. Thus, significant management attention must be paid to the development of the PMB, as well as to progress measurements, in order to ensure the accuracy and utility of EVM data.

Development of the PMB begins with the contract work breakdown structure (WBS). The contractor decomposes WBS elements into individual "work packages," which are relatively short-duration, discrete tasks with (see Fig. 12.1) an identified product or outcome, an assigned start and finish date, a cost budget to accomplish the work, and an assigned office or individual with responsibility to accomplish the work.

For contracts with longer durations, it is usually not possible to accomplish detailed work-package decomposition for the entire POP. In such cases, the budgets and durations for future tasks (e.g., a missile flight-test phase that follows a three-year period of missile design) are held in "planning packages,"

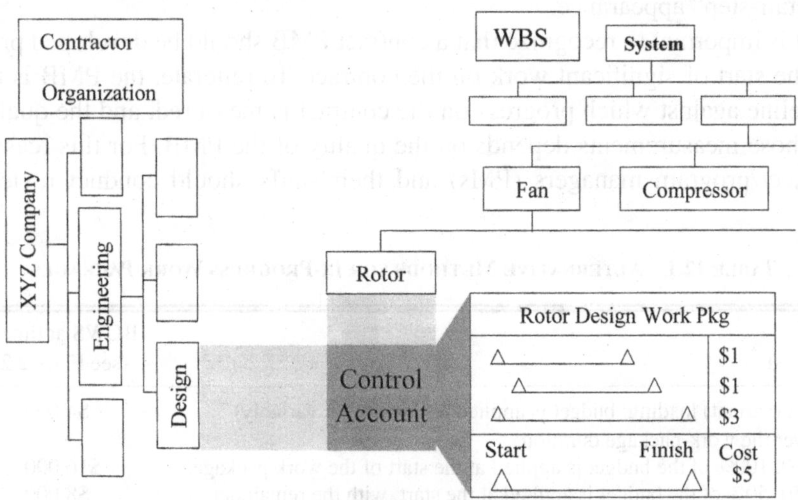

Fig. 12.1 Contract work packages and the contractor's organization.

and the decomposition to work packages is deferred until later in the POP when the details of required work can be determined more accurately.

WBS work packages are the building blocks for the PMB because they reflect the following:

- *Performance*: work with an associated product or outcome
- *Cost*: the assigned budget for that work
- *Schedule*: the start and finish dates for the work

The PMB is a "roll-up" of contract work packages, thus representing a plan for budget application over the POP. Each point on the PMB is a *budgeted cost of work scheduled* (BCWS) (also referred to as *planned value*) at a given time.

12.4 MEASURING WORK IN THE PMB

The PMB explicitly shows the planned costs of the contract relative to the schedule. Performance is not explicitly depicted, but rather is implicitly shown by the projected progress of work package completion.

Determining how progress is to be planned for an individual work package after its start, but before its completion, requires assumptions to be made concerning the application of budget over the duration of the work package. Some common ways of planning for the application of budget for a work package are listed in Table 12.1.

Notice that BCWS—and thus the shape of the PMB when graphed—will vary according to the method selected. Figure 12.3 depicts a PMB for a contract with six level- or variant-loaded work packages. If the work packages followed the 0–100, the 50–50, or the 100–0 methods, the PMB would have a "stair-step" appearance.

It is important to recognize that a contract PMB should be developed prior to the start of significant work on the contract. To reiterate, the PMB is the baseline against which progress on the contract is measured, and the quality of those measurements depends on the quality of the PMB. For this reason, project/program managers (PMs) and their staffs should conduct reviews

TABLE 12.1 ALTERNATIVE METHODS FOR IN-PROGRESS WORK PACKAGES

Method	BCWS at time T (see Fig. 12.2)
Level (variant) loading: budget is applied uniformly (or variably) over the work-package duration.	$4,000
100–0: 100% of the budget is applied at the start of the work package.	$16,000
50–50: 50% of the budget is applied at the start, with the remainder applied when the work package is completed.	$8,000
0–100: 100% of the budget is applied when the work package is completed.	$0

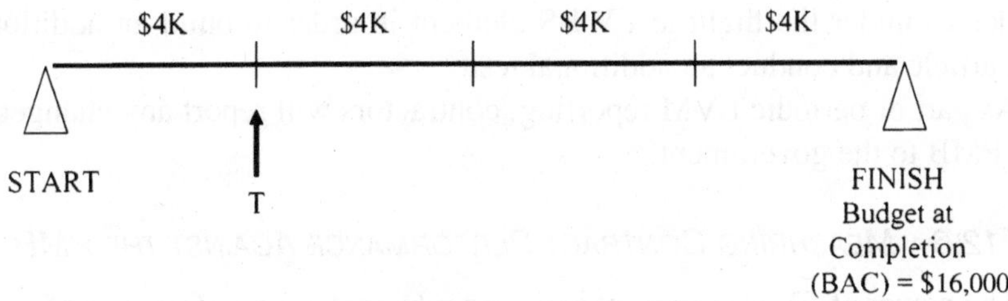

Fig. 12.2 Notional work package.

(called integrated baseline reviews) with the contractor soon after contract award in order to ensure the quality of detailed work package scheduling and budgeting. This review provides all stakeholders with a thorough understanding of their roles in the project as well as its goals, plans, and risks—thus helping ensure buy-in from all concerned.

12.5 CHANGES TO THE PMB DURING CONTRACT PERFORMANCE

Changes to the PMB can occur during the POP because of several circumstances:

1) Contract scope can change, with either additions or deletions to work packages, along with their corresponding schedules and budgets.

2) Start and/or end dates of work packages can shift, with the total scope of work remaining constant. This would cause a change in the PMB's shape, but the budget at completion (BAC) would not change.

3) *Management reserve* (MR) can be applied to WBS elements and work packages. MR is a separate, budgeted amount in the contract total cost, which is set aside for unforeseen contingencies. For example, in the event of a missile flight-test failure, some MR can be added to the budgets of work

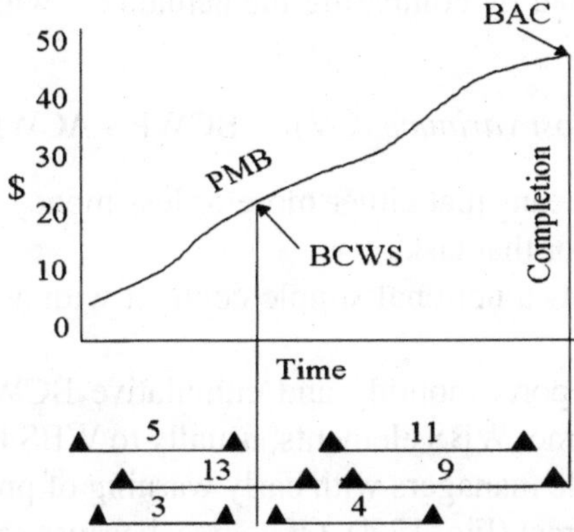

Fig. 12.3 PMB for a contract with six work packages.

packages under the flight-test WBS element in order to build an additional test article and conduct an additional test.

As part of periodic EVM reporting, contractors will report any changes to the PMB to the government.

12.6 MEASURING CONTRACT PERFORMANCE AGAINST THE PMB

The power of EVM comes from the term *budgeted cost of work performed* (BCWP), also called *earned value*. BCWP measures the value of the work performed on a task in terms of the budget for that work. Whereas BCWS measures the value of the planned or scheduled work, BCWP measures the value of the actual or earned work.

BCWP provides the basis for determining variances from the plan. Referring to Fig. 12.2, and using the 50–50 method for measuring work in progress, the BCWS projected at time T is $8,000. BCWP (the earned value), once time T actually arrives, will be one of the following:

- $0—if work on the work package has not yet started
- $8,000—if work has begun but is not yet complete
- $16,000—if work on the work package is complete

Thus, depending on how much work has actually been accomplished, the work package might have a negative (unfavorable) variance, no variance, or a positive (favorable) variance.

Schedule variance (SV) = BCWP – BCWS

Note this is a schedule (temporal) variance that is expressed in budgetary terms. SV indicates whether contract tasks are on, behind, or ahead of schedule.

To determine contract cost performance, another term is needed: *actual cost of work performed* (ACWP). ACWP for a work package is determined by the true costs (e.g., labor, materials, etc.) of performing the work. Cost variance is determined by comparing the actual cost with the budgeted cost of work performed:

Cost variance (CV) = BCWP – ACWP

A cost variance means that either more or less money was spent on a task than was budgeted for that task.

Figure 12.4 depicts a notional simple contract with associated EVM data and variances.

The contractor reports monthly and cumulative BCWS, BCWP, ACWP, CV, and SV for contract WBS elements, usually to WBS level 3. Unfavorable variances can provide managers with early warning of problems surfacing in any parts of the contract (Fig. 12.4). Of course, because several positive WBS element variances could mask one significantly unfavorable variance, it is

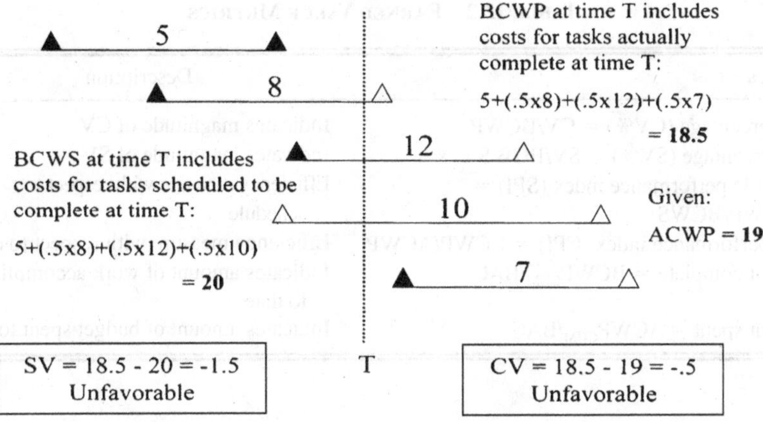

Fig. 12.4 Contract EVM data and variances (assumes the 50–50 method for measuring work in progress).

important for managers to examine the details of EVM data and to not simply skim summary data (see Fig. 12.5).

12.7 EVM ANALYSIS

The simple metrics BCWS, BCWP, ACWP, CV, and SV provide the foundation for other metrics for more sophisticated EVM analysis. Some of the more common metrics are included in Table 12.2.

EVM analysis identifies the root causes of observed variances. A "one-time" problem event can cause a single reported unfavorable variance, whereas systemic process problems will result in increasingly unfavorable variance. Once the causes of variances are identified, corrective actions can be implemented.

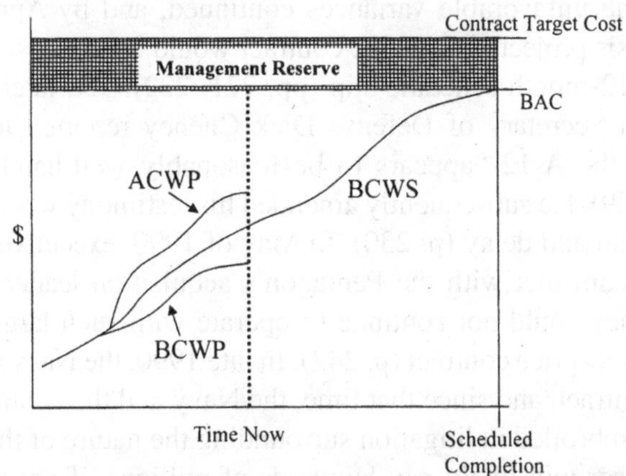

Fig. 12.5 Cumulative graphical EVM data for a contract with unfavorable schedule and cost variances.

TABLE 12.2 EARNED VALUE METRICS

Metrics	Description
CV percentage (CV%) = CV/BCWP	Indicates magnitude of CV
SV percentage (SV%) = SV/BCWS	Indicates magnitude of SV
Schedule performance index (SPI) = BCWP/BCWS	Efficiency measure with respect to schedule
Cost performance index (CPI) = BCWP/ACWP	Efficiency measure with respect to cost
Percent complete = $BCWP_{CUM}$/BAC	Indicates amount of work accomplished to date
Percent spent = $ACWP_{CUM}$/BAC	Indicates amount of budget spent to date

12.8 PROJECTING FUTURE CONTRACT PERFORMANCE

EVM is a powerful method for forecasting future contract performance (e.g., cost and schedule growth) based on past performance (see Box 12.2). Research has shown that EVM enables exceptionally accurate estimates of future performance on a contract after it is as little as 15% complete [3].

BOX 12.2 EVM AND THE NAVY'S A-12 [4]

The book *The $5 Billion Misunderstanding: The Collapse of the Navy's A-12 Stealth Bomber Program* recounts the managerial failures of the A-12 program leadership when faced with unfavorable EVM information. From December 1988 to February 1989, program analysts reported increasing unfavorable variances, but supervisors took no action, believing the contract's fixed-price structure protected the Navy's interests. The unfavorable variances continued, and by April of 1990 EVM analysis projected the A-12 contract would have a cost overrun of $1B and a 12-month schedule slip (pp. 221, 222). Yet, later that same month, then Secretary of Defense Dick Cheney reported to the U.S. Senate that the A-12 "appears to be reasonably well-handled at this point" (p. 229). He subsequently amended his testimony when informed of the overrun and delay (p. 230). In May of 1990, executives from the contractor team met with the Pentagon's acquisition leaders to notify them that they could not continue to operate with such large overruns under the fixed-price contract (p. 242). In late 1990, the Navy terminated the A-12 contract, and since that time, the Navy and the contractor team have been embroiled in litigation surrounding the nature of the termination. The loser will likely pay hundreds of millions, if not billions, of dollars in penalties, interest, and court costs.

TABLE 12.3 METHODS FOR DETERMINING EAC

EAC =	Assumes
$BAC + CV_{CUM}$	No change in CV over the remainder of the POP
BAC/CPI_{CUM}	Entire contract will reflect the cost efficiency to date
$ACWP_{CUM} + [(BAC - BCWP_{CUM})/CPI_{CUM}SPI_{CUM}]$	Remaining work will reflect both cost and schedule efficiencies experienced to date

An item of particular interest to acquisition managers is a contract's estimate at completion (EAC) based on performance to date. Obviously, at the start of a contract, EAC = BAC. After work begins on the contract, however, EAC should be adjusted to account for actual performance to date. Numerous methods exist for obtaining EAC, and each manager should select the method he or she feels is most appropriate for the unique circumstances of the contract under analysis. Three simple methods appear in Table 12.3.

The realism of any EAC can be tested using the To Complete Performance Index (TCPI), a metric for the efficiency with which the contractor must perform in order to achieve the EAC:

$$
\begin{aligned}
TCPI_{EAC} &= \text{Work Remaining/Cost Remaining} \\
&= (BAC - BCWP_{CUM})/(EAC - ACWP_{CUM})
\end{aligned}
$$

Clearly, if $TCPI_{EAC}$ is greater than one while CPI_{CUM} is less than one, the contractor must exhibit substantial improvement over the remainder of the POP in order to meet that EAC. If such improvement is unlikely, then that EAC is unrealistic.

12.9 CONCLUSION

This chapter has given a brief overview of the basics of EVM in order to highlight its usefulness as a managerial tool. Much more sophisticated and in-depth analysis is possible using the basic elements and techniques just presented. Additionally, a variety of software tools is available to aid PMs in decision making by handling EVM analysis and by presenting its results in easily understood graphical formats.

REFERENCES

[1] Government Accountability Office, "Defense Acquisitions: Assessments of Selected Weapon Programs," GAO-07-406SP, Washington, D.C., March 2007.

[2] General Accounting Office, "Defense Acquisitions: Despite Restructuring, SBIRS High Program Remains at Risk of Cost and Schedule Overruns," GAO-04-48, Washington, D.C., Oct. 2003, p. 12.

[3] Fleming, Q., and Koppelman, J., *Earned Value Project Management*, 3rd ed., Project Management Inst., Newtown Square, PA, 2006.

[4] Stevenson, J., *The $5 Billion Misunderstanding: The Collapse of the Navy's A-12 Stealth Bomber Program*, Naval Inst. Press, Annapolis, MD, 2001, p. 238.

STUDY QUESTIONS

12.1 Why is EVM not typically used on fixed-price or level-of-effort contracts?

12.2 How does the performance measurement baseline reflect contract cost, schedule, and performance?

12.3 Describe the purpose of an integrated baseline review.

12.4 Why is the selection of a method to measure work in progress in a work package important?

12.5 What are some reasons why the PMB might change during a contract's period of performance?

12.6 How might unfavorable variances be "masked" in EVM reporting?

12.7 In addition to the EAC formulae presented in this chapter, what other equations could be developed, and under what conditions might they be useful?

Chapter 13

STRATEGIC PURCHASING

BRYAN HUDGENS*
NAVAL POSTGRADUATE SCHOOL, MONTEREY, CALIFORNIA

LEARNING OBJECTIVES

- Appreciate the role of purchasing as an element of organizational strategy
- Comprehend how strategic purchasing contributes to competitive advantage
- Comprehend strategic factors involved in outsourcing organizational functions
- Appreciate the role of purchasing within the context of supply-chain management
- Comprehend the importance of supplier relationships and supply-base management
- Appreciate various strategic analysis tools relevant to strategic sourcing
- Appreciate factors influencing the organization of a strategic purchasing function

13.1 INTRODUCTION

Purchasing has evolved from a historically clerical function to one that supports and influences organizational strategy (see [1, 2]). This chapter provides an overview of how purchasing achieves these new goals. It begins with a brief review of purchasing and how it has evolved into a more strategic function. From there, the chapter discusses the notion of competitive advantage and how the purchasing function might contribute to an organization's competitive advantage. Proceeding from the discussion of competitive advantage, the chapter next provides an overview of strategic considerations involved in

*Lecturer, Graduate School of Business and Public Policy.

the decision to outsource and presents several concepts that are helpful in sourcing goods and services strategically. The increased emphasis on competitive advantage and outsourcing leads to a discussion of supply-chain management. A critical aspect of strategic sourcing is the supply base itself, and the chapter next explains some core considerations for managing the supply base and improving relationships with key suppliers. The chapter concludes by discussing recent approaches to organizing the strategic purchasing function.

13.2 Purchasing and Organizational Strategy

Purchasing has not always been considered a strategic function; historically, it has been viewed as a clerical function serving simply to acquire whatever the organization needs [1, 2]. In this chapter, consistent with other references, the term *purchasing* will refer to this basic functional definition of acquiring an organization's goods and services. More recently, however, many organizations have benefited from buying strategically. Specifically, organizations that take a more strategic approach to purchasing have enjoyed improvements of 20% or more in cost, schedule, and quality [2].

The Office of Management and Budget has defined strategic sourcing as a "collaborative and structured process of critically analyzing an organization's spending and using this information to make business decisions about acquiring commodities and services more effectively and efficiently" (data available online at http://www.whitehouse.gov/omb/procurement/comp-src/implementing strategic-sourcing.pdf. Although organization- or enterprise-level goals of cost reduction and mission improvement are included in this definition, this chapter introduces a third term, *strategic purchasing*, to emphasize the ability of purchasing to align with and influence enterprise-level strategic goals [3]. (For example, see Defense Acquisition University's strategic sourcing training course [4], which suggests—correctly—that organizations can source strategically at any organizational level. Strategic purchasing includes the notion of strategic sourcing, which can operate below the enterprise level; however, it also emphasizes the ability of purchasing to align with and influence enterprise strategy.) The next section elaborates on this idea in greater detail.

13.3 Strategic Purchasing and Organizational Competitive Advantage

This chapter suggests that purchasing can contribute to the achievement of enterprise-level strategic goals. This section defines the concept of competitive advantage, presents models that an organization can use to assess its competitive advantages, and suggests ways in which purchasing can use these models to contribute to organizational strategy. A complete discussion of organizational strategy is obviously well beyond this text. This chapter takes a very simplified approach to understanding organizational strategy by analyzing industries and individual firms and discussing an analytic tool useful for each level of analysis; specifically, Michael Porter's Five Forces Model analyzes the competitive intensity within an industry, whereas the Resource Based View of the Firm and the related concept of an organization's core competencies analyze the competitive ability of individual organizations. Why does this matter? Simply, organizations can apply these tools to become more competitive. Although this material is based on corporate strategy, this chapter suggests that purchasing organizations—including those within the federal government and, specifically, the Department of Defense—can use these tools to source strategically and contribute to enterprise-level goals.

13.3.1 INDUSTRY ANALYSIS (PORTER'S FIVE FORCES)

Industry-level analysis assumes that the economic structure of industries (i.e., whether they are monopolistic, perfectly competitive, or somewhere in between) influences how firms within those industries behave (e.g., whether they take prices from the market or try to differentiate themselves from competitors) and, thus, influences their ultimate profitability. This discussion follows any of several standard texts (e.g., [5, 6]). Perhaps obviously, Michael Porter's classic [7] is the starting point for understanding his model. Michael Porter's Five Forces model provides a framework for understanding forces influencing the overall attractiveness of an industry (in terms of average profitability) when compared with other industries (Fig. 13.1). Porter's Five Forces include the intensity of competition (rivalry) within the industry itself, the likelihood of new competitors entering the industry, the presence of

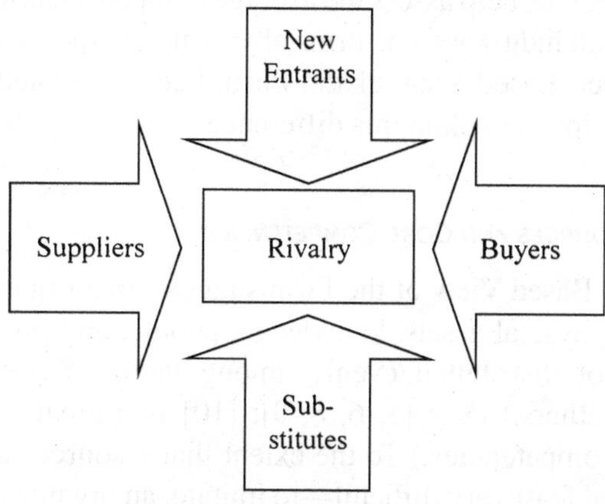

Fig. 13.1 Five Forces Model of industry analysis (adapted from [7]).

substitute products that might meet the same need in a different way, and the relative power of customers and suppliers vis-à-vis firms in the industry.

Rivalry, the intensity with which firms in an industry already compete among themselves, depends on several factors—including the number and size of the firms and their ability to differentiate themselves and add production capacity (the more costly to do so, the greater the rivalry). New firms will try to enter an economically attractive industry to earn higher profits; firms already in the industry will resist their attempts through a variety of means, known collectively as barriers to new firms' entry. New entrants are less likely to enter an otherwise attractive industry when existing firms already have achieved economies of scale in their production capabilities or have other cost advantages (access to proprietary knowledge, technology or essential materials or even simple learning-curve efficiencies) or have differentiated their products sufficiently to achieve brand loyalty, signal their willingness to fight new entrants (e.g., with a price war), or when government tariffs and trade barriers exist. Substitute products reduce the attractiveness of an industry by effectively capping the profitability of that industry; at some level of profitability, consumers will look to other sources to meet their needs (e.g., schedule videoconferences instead of flying to meet in person). Finally, buyers and suppliers reduce the level of industry profitability by demanding price reductions (buyers) or by raising prices (suppliers). Buyers can threaten industry profitability by demanding price reductions most effectively when there are few buyers, when the products they need are standard, or when they can credibly threaten to enter the industry and compete directly with their supplier ("backward integration" within the supply chain; see Sec. 13.5 for a discussion of supply chains). Suppliers operate in much the same way: utilizing fewer suppliers, providing essential components, or possessing a credible ability to enter the industry and compete ("forward integration") can increase their prices, thus reducing industry profits.

Porter's Five Forces help assess the average competitiveness of an industry; however, within an industry some firms obviously prosper, whereas others do not. The Resource Based View of the Firm, and the related notion of core competencies, helps to explain this difference.

13.3.2 FIRM RESOURCES AND CORE COMPETENCIES

The Resource Based View of the Firm suggests that organizations control resources (e.g., physical assets, knowledge, people, and processes), but that resources are not distributed evenly among them. (Some have different resources from others.) (See [5, 6, 8, 9]; [10] is a great starting point for exploring core competencies.) To the extent that resources are different and impossible—or at least very difficult—to imitate, an organization can employ those resources to gain an advantage. Organizations can create an advantage to the extent their resources are valuable, rare, inimitable, and to the extent the organization organizes itself to exploit them. The greater the extent to which all four factors are present, the greater the advantage; indeed, if all four factors are present, the organization can enjoy a sustained competitive advantage, meaning it can employ its resources to create value in a way essentially no other firm can match. Organizations can build on their resources to create *capabilities* (essentially processes performed very well) and *competencies* (essentially capabilities performed very well). An organization's competencies can be considered core competencies; when the organization performs them well and others do not, the competencies contribute significantly to the value the customer believes it receives. In addition, the competencies span multiple product lines, thus providing access to many markets.

13.3.3 INDUSTRY- AND FIRM-LEVEL ANALYSIS FOR DEFENSE CONTRACTING

Corporate strategy experts suggest that when a firm understands its industry and its resources and core competencies it will be able to compete more effectively. These concepts can also be helpful to organizations attempting to source strategically. Although such activities are not part of the traditional purchasing process, organizations that take the time to understand the industry from which they are purchasing, and the strengths (and weaknesses) of the major firms within it, can be better prepared to make well-informed sourcing decisions to purchase from the best-available suppliers. For example, consider a contracting office that awards a contract to a firm that is relatively dependent on powerful suppliers for the products it provides the government; understanding the industry power structure can help a contracting office better structure its contract. On the one hand, it is true the government's supplier is obligated to provide the products it agreed to provide under the contract; on the other hand, the government should be aware of the relative risks involved in dealing with a supplier that is itself dependent on its suppliers to fulfill the contract. If the government's contractor encounters problems with its suppliers, the government is ultimately hurt when it does not receive needed products. Careful industry analysis would have helped the government to understand this situation and take steps to manage the risks. The next section builds on the notion of core competencies by discussing critical factors that an organization considers when assessing whether or not to outsource.

13.4 STRATEGIC OUTSOURCING CONSIDERATIONS

Outsourcing is the "buy" half of what is generally considered the "make or buy" decision, in which an organization determines whether to produce a good or service or procure that good or service from another organization. Outsourcing decisions require careful consideration of several factors; this section discusses three critical strategic considerations, including core competencies, production capability, and total cost.

Core competencies are those that are important to the organization, that the organization does well, and that create value for the organization by creating value for many customers in many markets. Organizations do not, by definition, outsource recognized core competencies; however, rather than enhancing organizational performance, some former core competencies can begin to restrict performance. (See [11] for a good discussion of these so-called core rigidities.) The organization should stop doing these activities. In some cases, the former core competence might add no value and should not be done at all. It is possible, however, that, although the organization should not perform the activity that formerly was a core competence, that activity might still be necessary in some sense. In these cases, the organization could outsource the activity (ideally to an organization for which that activity is a core competence).

Most outsourcing decisions do not involve core competencies, but rather supporting or enabling competencies. For the military, outsourcing combat capability is generally not an option, whereas outsourcing the housekeeping function in on-base visiting quarters is reasonable. (Despite the growth of private military companies, the nation's combat capability remains by and large with the military.) The decision whether to outsource supporting or enabling competencies such as medical services or procurement can be more difficult. Such capabilities are important to the military's overall mission, and decisions to outsource these and similar functions should be made very carefully. [This is not an either-or situation; in many cases, portions of supporting competencies can be outsourced. A typical example is to outsource some medical services at a home installation (e.g., garrison) while maintaining some military providers, who are capable of deploying forward.] Finally, outsourced competencies are generally considered unable to provide a long-term competitive advantage because other organizations can presumably outsource them to the same source, thus duplicating the competency.

Production capacity is a second critical consideration with outsourcing. Outsourcing production of a good or service leads to several risks, including the loss of control of an important process, and "hollowing out," or losing the ability to perform the internally function if needed. On the supplier's side, however, production capacity might not be available, or might be available only at the cost of adding a second (or third) production shift—a potentially expensive and possibly prohibitive proposition. Another consideration is whether any one supplier has the available capacity to produce the capability required; if not, multiple sources will be needed.

Finally, total cost considerations should weigh heavily on the decision to outsource. The calculation of the in-house total costs with which outsourced costs will be compared can be problematic. In-house costs include such potentially difficult issues as overhead allocation and costs of capital and quality. Overhead can be particularly vexing because overhead costs can be allocated in several different ways, leading to significantly different costs for a given product or service.

Although not the only three considerations involved in outsourcing—as suggested, another critical consideration beyond the strict historical purview of purchasing would be the impact on the organization's personnel—these three factors can help the organization make smarter strategic decisions about what products and services to outsource. As organizations outsource noncore competencies, they form relationships with other organizations, which in turn have relationships with still other organizations. The next section discusses some implications of these interorganizational relationships.

13.5 PURCHASING AND SUPPLY-CHAIN MANAGEMENT

Preceding sections discussed how an organization decides to compete within economically attractive industries. If the notion of a core competence has any validity, then organizations should (must) seek out other organizations with complementary and supplementary competencies in order to maximize their ability to deliver value to their customers. The chapter has also discussed various strategic considerations involved in the outsourcing decision. This section elaborates on the idea of these interorganizational relationships, asking what kind of relationships organizations should have with other organizations and under what conditions (Many good sources discuss relational contracting; see, especially, [12]). From there, this section suggests that organizations exist in a chain (or really a web) of relationships they maintain with other organizations. These interrelationships have come to be called supply chains, which represent the interconnection of firms that extends from the extraction of raw materials, to utilization by the end users, to ultimate disposal of a product at the end of its useful life [12, 13]. Recognizing the implications of these interrelationships and managing them effectively is vital to organization competitiveness. Some experts suggest supply chains have replaced individual organizations as the new unit of competition [14].

The trend in supplier relations is toward just that—relationships. In the past, firms operated at arms' length, managing transactional relationships characterized by a win-lose philosophy that showed little concern for others and that were formed based on price and were dissolved upon completion of the immediate task; no promise of future business was implied or assumed. More recently, firms have discovered that, although certainly not panaceas, closer relationships (usually with fewer suppliers) offer significant benefits. The idea of fewer suppliers, and how to deal with them, will be discussed in greater detail later. For now, the primary point is that organizations now focus more on their own core competencies and choose to work, often very closely, with other organizations who can supplement their competencies.

As organizations begin to work with each other, they begin to form a *supply chain*: three or more organizations linked directly by reciprocal flows of products and/or services, information, and money [12, 13]. These supply chains can ultimately reach backward to the raw materials and forward to the final end-use customer and on to the disposal of the products at the end of their useful lives (see Fig. 13.2) [13, 15]. Essentially all organizations operate in supply chains.

Supply management describes the process of managing this interlinked chain of organizations. Some organizations, however, truly understand the strategic implications of supply chains; these organizations have a supply-chain orientation. Supply-chain management is important for many reasons, not the least of which is it offers the potential to achieve multiple synergistic goals of improving service (shorter cycle times and greater reliability), increasing revenue (by working with partners to optimize the entire supply chain), and improving satisfaction and loyalty among supply-chain partners and customers. This creates the possibility of a virtuous circle, in which increasingly better service leads loyal customers to give repeat business to a supply chain that is operating more and more efficiently, thus gaining greater profits from increasing revenue.

What are the implications of supply-chain management to the (strategic) purchasing function? Perhaps most obviously, purchasing is responsible for managing the relationships with all suppliers to the organization. Thus, it is responsible for the entire "upstream" portion of the supply chain—all of the organizations between the buying organization and the raw material.

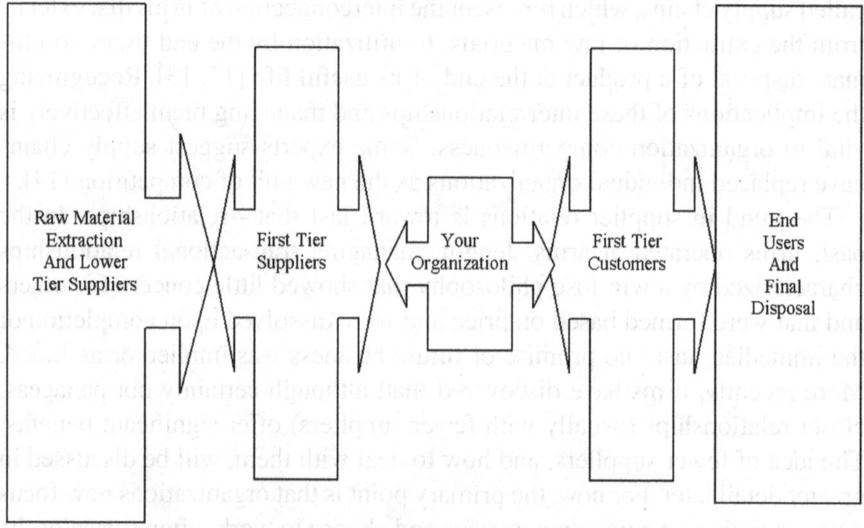

Fig. 13.2 General supply-chain model (adapted from [13, 15]).

Additionally, purchasing is only one of the functions involved in supply-chain management (just a few of the others include transportation, inventory management, production, and customer service). However, the nature of supply chains is that each function affects the others, so that the supply chain must be managed as a whole. Failure to do so can lead to suboptimization, whereby one function performs very well, but, in doing so, its effectiveness causes other functions to underperform.

For example, one of the easiest ways for a purchasing function to appear very effective is to buy in large quantities; all else being equal, this results in lower unit costs. (This is essentially the business case for warehouse-buying centers like Costco and Sam's Club.) Sometimes this makes sense; everyone knows that the best time to buy wrapping paper is shortly after Christmas. Similar "good deals" are available to organizations for various reasons, whether seasonality, sales, market gluts, or other reasons. Sometimes organizations should buy as much of what they need as possible to take advantage of low prices. On the other hand, in many cases, these good deals are illusory. What an organization buys, it typically must store; these costs can be nontrivial. This is the idea of the economic order quantity, which seeks to balance ordering and storage costs (for example, see [12]). Organizations also usually have to move what they buy, and transportation can be expensive (not always, but sometimes) or slower. (Large quantities bought overseas often require seagoing transport rather than the much faster air shipment.) In the case of just-in-time production, a large purchase of material can cause huge problems for production bottlenecks. The point of these examples is that supply chains require careful management of the entire function, and purchasing must sometimes suboptimize its own performance (pass up the "deal of the century") to ensure optimal performance of the entire supply chain.

Whether or not the supply chain is the new unit of competition, supply chains and suppliers are undeniably important. The next section of this chapter explores how an organization can better select suppliers and manage the entire supply base to improve its performance.

13.6 SUPPLIER SELECTION AND SUPPLY-BASE MANAGEMENT

As organizations outsource more, suppliers take on greater importance in terms of their impact on bottom-line mission performance. The good news is that organizations can improve significantly their overall performance by managing suppliers effectively. A complete discussion of source selection, supply-base management, and supplier development would require at least one book. (See the recommended reading for some places to start.) However, in keeping with its strategic theme, this chapter will suggest strategic approaches to supplier selection and supply-base management. This section addresses the selection of good suppliers to form a supply base, the effective management of that supply base, and the development of both individual suppliers and the entire supply base to greater capability in the hopes that their greater capability will improve the buying organization's performance [12]. It begins with a discussion of how organizations can develop souring strategies for products or services.

13.6.1 SOURCING STRATEGY DEVELOPMENT

Portfolio analysis is a simple but powerful approach to strategic sourcing. (See [12] for a good overview of portfolio analysis tools.) Portfolio analysis techniques assess purchases—or the firms involved in the purchase—by, typically, two major characteristics. By cross-indexing these two characteristics, the organization can categorize purchases into four broad categories, each of which provides insights that can lead to more strategic purchasing decisions. For example, purchases might be categorized according to any number of characteristics, including the complexity of the market, the buyer's relative power in the market, the cost or value (importance) of the purchase to the buying organization, the total cost (total spend) of the product or service category, or the acquisition risk, or by any other characteristics that make sense [12, 16, 17]. Drawing from [12 and 16], this section presents a sample portfolio analysis by market complexity and purchase value, and discusses how the enterprise can use it to make wiser strategic buying decisions. Analyses using other characteristics work similarly.

Categorizing purchases by market complexity and value of the purchase suggests four broad strategies. Some items are not terribly costly and the market for them is relatively simple, consisting of standardized goods and services available plentifully from many suppliers; these items present little risk to the buying firm should supplies be disrupted. In such cases, the purchasing organization would want to emphasize efficiency. A useful strategy for all products and services meeting this general description is to streamline the purchasing process. For example, the purchasing organization might choose to issue purchasing credit cards to appropriate personnel or put in place master contracts with ordering officials distributed throughout the enterprise (perhaps even through online ordering portals).

When the organization spends a comparatively large amount of money on items available from a noncomplex market, it should consider consolidating its purchases with a single or very few suppliers to attain economies of scale and gain leverage in the market. Essentially, these cases enable the organization to exploit its relative power in the market.

Critical items represent potential "headaches" because, while the organization does not spend much on them relatively speaking, they can cause disproportionate problems should their availability be disrupted. One major goal for the purchasing organization is to find ways to avoid the problems altogether;

perhaps the products can be redesigned to enable more suppliers to compete for business. Otherwise, the purchasing organization should take steps to ensure adequate contingency plans are in place if (when) problems arise; organizations might decide to work more closely with their suppliers to catch problems early, and to carry extra inventory to mitigate consequences of problems that do arise.

Finally, strategic items represent high-dollar expenditures in complex markets. Such items should be managed specially. Many organizations, including the government and the DoD, set up specific program offices to manage such purchases, and these offices often employ the full spectrum of strategic purchasing techniques.

13.6.2 SUPPLIER SELECTION

Once a sourcing strategy is in place, the organization can identify, evaluate, and select one or more suppliers consistent with that strategy. Identifying potential suppliers is perhaps less difficult than it was a few decades ago; purchasing experts can find them through the ubiquitous internet, colleagues (including in-house technical experts), industry groups, trade shows, trade publications, and even other suppliers. Given the array of sources available to identify suppliers, the buying office frequently will confront multiple offers for any given purchase request.

Faced with this array of potential suppliers, the buying office often considers winnowing the list of suppliers. The buyer might determine whether offerors are responsible, that is, capable of performing the task, both technically and financially. Technical expert opinions and financial data from both the offeror and published sources can determine responsibility. A second approach to narrowing the field of offerors to those most likely to be viable is to identify a so-called relevant range, which eliminates offers that appear to be either too low or too high to have a reasonable chance of being selected. Offers that are too low might indicate the offeror does not adequately understand the buyer's needs (or is trying to "buy in" to the contract with an unrealistically low price); offers that are too high might indicate a similar lack of understanding of the requirement or a "gold-plating" approach to meeting it. In either case, the notion of a "relevant range" is that several offers appear to be able to perform the task, and their pricing falls within a range of offers that appears neither too high nor too low. Accordingly, offers outside this range can be disregarded.

Selecting the supplier(s) is often done in accordance with a predetermined list of criteria the purchasing office believes are important to the purchase; for example, the buyer might consider cost, technical capability, quality, management expertise, past performance, and other criteria. These criteria are often weighted, usually in terms of percentages, to indicate their relative importance to the selection decision; for example, cost might be weighted as

25% of the total decision, whereas quality might be 40%, with other criteria making up the other 35% of the decision. Each offer that remained after the field was narrowed in the previous stage is then evaluated independently against these criteria. Finally, the purchasing office can negotiate and execute a contract with as many sources as necessary to meet the requirement.

The actual source-selection process might not be particularly different within elite purchasing organizations than it is in less strategically oriented organizations. However, strategically focused purchasing organizations move beyond the management of individual purchases to the management of the entire supply base. The remainder of this section introduces this concept.

13.6.3 Supply-Base Management

Supply-base management involves managing the suppliers with which an organization considers doing business (for example, see [2]). It is not a static decision, in that suppliers might be added or subtracted over time. Supply-base optimization is important because, although strategic purchasing offers clear and compelling benefits (the 20% improvements to cost, quality, and schedule mentioned earlier), these benefits come at a cost to the buying organization, which must invest time and resources to develop relationships with its suppliers. Doing so is not typically possible with all available candidates, so that strategically focused purchasing organizations reduce the number of candidate suppliers through one of several rigorous processes.

One possible supply-base optimization approach is to employ the well-known Pareto rule, which suggests that 20% of an organization's suppliers provide 80% of the organization's products and services; conversely, 20% (ideally, a different 20%!) cause 80% of the problems. To the extent that this idea is at least somewhat true, it provides a quick way to begin thinking about how to optimize the supply base. A related approach triages suppliers into those performing well already (keep them), those apparently incapable or unwilling to perform well (why keep them?), and those who, with some attention, could perform well (develop them). The Pareto and triage approaches are based on past performance. Other methods build on these approaches by adding a future-based criterion such as quality improvement. These approaches set future performance goals for suppliers, and those suppliers willing and able to meet these goals remain in the supply base.

Once suppliers, and a supply base, are chosen, the strategic purchasing organization can turn its attention to supplier development, which discipline argues that by improving an organization's suppliers, the organization itself wins.

13.6.4 Supplier Development

In manufacturing, approximately 55 cents of every dollar of revenue is spent on purchased goods and services [2]. The business case for supplier development is that the investment of some resources in improving suppliers offers the potential for disproportionate returns; $1 of costs saved by improving supplier performance equals $5–10 of revenue (assuming a profit rate of 10–20%). Beyond the purely cost-driven argument, as organizations outsource more, their ability to compete in the global arena becomes increasingly dependent on suppliers. The careful selection of and development of suppliers—and an entire supply base—offers the potential for great rewards. The strategic purchasing organization can improve even further by using strategic analysis tools such as spending analysis, total cost analysis, and value analysis. The next section introduces these additional strategic analysis tools.

13.7 Strategic Analysis Tools

Many experts would suggest this is where the strategic sourcing process actually starts—with an understanding of the requirements and organizational spending that strategic analysis techniques can contribute. (These experts are not wrong; however, this chapter emphasizes the strategic influence of the purchasing function and thus starts with that broader theme.) This section presents an overview of several such techniques that help an organization answer how much the organization spends, on what, and with whom? (See [12] for a good introduction of each technique.) Spend analysis gives a historical view of an organization's spending, providing data from which to make strategic decisions on future spending. When this analysis is combined with a decent forecast of future requirements, an organization can be better prepared to understand its role in the market and make the best possible business arrangements. Unfortunately, the initial purchase price is only one component of what products actually cost in the long run; very often that long-run cost is by far the largest component of the total ownership cost. Assuming the organization has an accurate total cost estimate, how then can it improve its total cost? One way is to redesign components using value analysis, which seeks to maximize the "bang for buck" in a product's design. This section briefly introduces these valuable tools: spend analysis, total cost analysis, and value analysis. Used together, these tools can help an enterprise purchase strategically.

Spend analysis seeks to analyze both purchases and the supply base to understand risks and opportunities involved in how—and from whom—the organization buys [18]. Opportunities arise when suppliers have multiple contracts; products are purchased from multiple suppliers; many offices within the enterprise are purchasing the same items; or needed items are experiencing cost growth. The enterprise might benefit in these cases by consolidating its purchases among buying offices, suppliers, and contracts. Conversely, spend analysis might identify risks including single sources, lack of incentives to suppliers, low demand, and other factors. In these cases, the enterprise becomes aware of these risks and can seek ways to manage them.

Total cost analysis seeks to account for all costs involved in acquiring, using, and disposing of a product over its life cycle. This can be a very labor-intensive task (consider trying to do life-cycle costing for an airplane); therefore, it is likely most appropriate for more expensive, complicated items. Because the analysis can be complicated, experts from any appropriate functional discipline (e.g., engineering, logistics, finance, purchasing, program management, cost/price analysis) should participate When appropriate, however, total cost analysis can help an organization avoid the trap of an attractive purchase price that brings with it prohibitively expensive operating and maintenance costs.

Value analysis involves reviewing the design of a product and, essentially, asking every component of that product to "buy its way" into the product in terms of value added. The goal is to reduce the cost of the product by ensuring every component of the product (or step of a service) adds value appropriate to its cost. High-cost components might be eliminated (if low value is added from the customer's perspective) or redesigned or replaced (if they add relatively high value).

Spend analysis seeks to understand how and from whom the organization buys and how much it spends. Total cost analysis emphasizes the criticality of understanding other costs incurred in addition to the purchase price. Finally, value analysis can help the enterprise maximize the value the customer gains from its products at the lowest possible production cost. Used together, these techniques can allow the enterprise to make much wiser purchasing decisions.

13.8 ORGANIZING FOR STRATEGIC PURCHASING

Having explored how a purchasing organization can contribute to enterprise-level strategic goals, and discussed how the purchasing organization can perform its functions more strategically, this chapter closes with a brief overview of how the purchasing function might organize itself most effectively to achieve these twin goals [2].

One initial consideration of an enterprise as it organizes the purchasing function should be the time-honored debate between centralization and decentralization. Should all members of the buying office be collocated, or should they be dispersed with internal customers throughout the organization? Centralization offers efficiencies in areas such as volume of purchases and collection of organization-wide procurement data; it also offers the purchasing professionals the ability to specialize in particular skill sets. Unfortunately, centralizing the purchasing function can cause customer service to falter—or at least can create the perception that it has or will falter. Decentralization of purchasing personnel can neutralize the perception of poor customer service; this happens in part by enabling the purchasing personnel to better understand the "mission" of the local organization they support. On the other hand, some of the advantages of centralization—volume efficiencies and specialization—are lacking. Thankfully, this choice is a false dilemma. Purchasing need not always adopt an all-or-nothing stance toward centralization or decentralization. Indeed, forward-thinking purchasing organizations have found ways to capitalize on the strengths while minimizing the weaknesses of both.

One new organization model for strategic purchasing that leverages the advantages of centralization while retaining some of the key advantages of decentralization is the commodity council. A commodity council is a permanent organization, staffed by a team of functional experts (including representatives from field organizations) tasked with developing and managing the acquisition strategy for a commodity. (Commodity is used here to mean any natural grouping of goods or services; e.g., IT support, medical support, and office support are all "commodities" in this sense.) This approach leverages the strengths of centralization—collecting organizational data and developing skills in strategic analysis—while enabling decentralized ordering against the agreement. The organization-wide purchasing agreement that results also typically allows exceptions for urgent local needs. These exceptions further diminish concerns from decentralized customers that they are "held hostage" by the centralized contracts administered elsewhere.

13.9 CONCLUSION

The purchasing function is (or at least should be) no longer a clerical function; rather, strategic purchasing offers organizations the possibility of significant improvement to cost, schedule, and quality. It can (and should) be viewed as a critical organizational function that supports and influences enterprise strategy. As we move into the 21st century, the opportunity to improve cost, schedule, and quality by 20% (or more) makes a compelling business case. As the Office of Management and Budget noted in its "call to arms" to the federal government's acquisition community, "The federal government spends approximately $300 billion on goods and services each year, and federal agencies are responsible for maximizing the value of each dollar spent. Therefore, agencies need to leverage spending to the maximum extent possible through strategic sourcing" (data available online at http://www.uspto.gov/web/offices/ac/comp/proc/OMBmemo.pdf).

13.10 ADDITIONAL RESOURCES

A good source to catalyze understanding of strategic purchasing is the *Supply Management Handbook* by Cavinato, Flynn, and Kauffman. This text provides a useful overview of many relevant topics. Another good source is the four-volume *Supply Management Knowledge Series* published by the Institute for Supply Management. Although expensive, many standard strategy textbooks can provide a good explanation of such topics as industry and firm-level analysis. (Porter's *Competitive Advantage* is self-explanatory as a useful source on his model.) Although also expensive, Monczka, Trent, and Handfield's text explains purchasing and supply-chain management well, as does Burt, Dobler, and Starling's. Although these textbooks are often a little dry and comparatively expensive, they provide the additional benefit of including further suggested reading for each topic they cover. For government (and especially DoD) employees, the Defense Acquisition University offers online courses that reinforce themes from this chapter. Finally, the Institute for Supply Management (www.ism.ws) provides many useful white papers, articles, and a good trade magazine on supply management topics. Membership is well worth the investment!

REFERENCES

[1] Leenders, M. R., Fearon, H. E., Flynn, A. E., and Johnson, P. F., *Purchasing and Supply Management*, 12th ed., McGraw-Hill, New York, 2002.

[2] Monczka, R., Trent, R., and Handfield, R., *Purchasing and Supply Chain Management*, 3rd ed., Thomson South-Western, Mason, OH, 2005.

[3] Ellram, L. M. and Carr, A., "Strategic Purchasing. A History and Review of the Literature," *International Journal of Purchasing and Materials Management*, Vol. 30, No. 2, 1994, pp. 10–18.

[4] Defense Acquisition Univ., Strategic Sourcing Training Course, CLC108 Strategic Sourcing Overview, 2007, https://learn.dau.mil/html/clc/clc.jsp [accessed June 2008].

[5] Oster, S. M., *Modern Competitive Analysis*, 3rd ed., Oxford Univ. Press, New York, 1999, Chap. 2, 3.

[6] Dess, G. G., and Lumpkin, G. T., *Strategic Management: Creating Competitive Advantages*, McGraw-Hill, New York, 2003, Chap. 2, 3.

[7] Porter, M., *Competitive Strategy*, The Free Press, New York, 1980.

[8] Barney, J., *Strategic Management and Competitive Advantage: Concepts*, 2nd ed., Prentice-Hall, Upper Saddle River, NJ, June 2007, Chap. 5.

[9] Barney, J., *Gaining and Sustaining Competitive Advantage*, 3rd ed., Prentice-Hall, Upper Saddle River, NJ, Nov. 2006.

[10] Prahalad, C. K., and Hamel, G., "The Core Competence of the Corporation," *Harvard Business Review*, Vol. 68, No. 3, 1990, pp. 79–91.

[11] Leonard, D., *Wellsprings of Knowledge: Building and Sustaining the Sources of Innovation*, Harvard Business School Press, Boston, MA, 1998, Chap. 2.

[12] Cavinato, J. L., Flynn, A. E., and Kauffman, R. G., *The Supply Management Handbook*, 7th ed., McGraw-Hill, New York, 2006.

[13] Mentzer, J. T. (ed), *Supply Chain Management*, Sage Publications, Thousand Oaks, CA, 2001, Chap. 1.

[14] Christopher, M., *Logistics and Supply Chain Management: Creating Value-Added Networks*, 3rd ed., Financial Times, Harlow, England, 2005, Chap. 10.

[15] Lambert, D., Cooper, M., and Pagh, J., "Supply Chain Management: Implementation Issues and Research Opportunities," *The International Journal of Logistics Management*, Vol. 9, No. 2, 1998, pp. 1–20.

[16] Kraljic, P., "Purchasing Must Become Supply Management," *Harvard Business Review*, Vol. 61, No. 5, 1983, pp. 109–117.
[17] Tang, C. S., "Supplier Relationship Map," *International Journal of Logistics*, Vol. 2, No. 1, 1999, pp. 39–56.
[18] Moore, N., Cook, C., Grammich, C., and Lindenblatt, C., "Using Spend Analysis to Help Identify Prospective Air Force Purchasing and Supply Management Initiatives," Rand Rept. DB-434-AF-13, RAND Corp., Santa Monica, CA, 2004, http://www.rand.org/pubs/documented_briefings/2004/ RAND_DB434.pdf [retrieved 2007].

STUDY QUESTIONS

13.1 As you accompany your installation commander to an unrelated meeting, he asks you to give him the CliffsNotes version of what strategic purchasing is and why it matters to him. Knowing you have only a few minutes, you realize your answer will be top level but also concise. What do you say?

13.2 Your new boss has asked you to take over a purchasing organization that operates functionally and tactically. She believes in the benefits of strategic purchasing and hired you to transform the organization. Outline broadly your plan of attack, in terms of what you and your new organization would need to learn and do, and how you might structure the organization effectively.

Chapter 14

ORGANIZATIONAL ASPECTS OF DEFENSE ACQUISITION

JOHN T. DILLARD*
NAVAL POSTGRADUATE SCHOOL, MONTEREY, CALIFORNIA

LEARNING OBJECTIVES

- Describe the decesion-making structure for acquisition projects
- Understand various ways in which program-management offices may be organized
- Describe the use of integrated product teams in acquisition projects
- Understand contributions of organizational theorists to defense acquisition
- Describe the relationships among project cost, control, and risk

14.1 INTRODUCTION

Defense acquisition is big business. The U.S. Department of Defense (DoD) routinely executes 12-figure budgets for research, development, procurement, and support of weapon systems. Acquisition is also a rule-intensive business. In addition to myriad laws governing federal acquisition, a plethora of regulations specify, often in great detail, how to accomplish the planning, review, execution, and oversight of acquisition programs. Because of these characteristics, the organizations responsible for DoD acquisition activities also tend to be large and rule intensive, reflecting the kinds of centralized, formalized, specialized, and oversight-intensive forms that correspond to the classic machine bureaucracy [1].

The bureaucratic nature of the DoD did not emerge recently, nor did it materialize by design. Rather, it reflects the cumulative accretion of laws, regulations, rules, and hierarchical levels over considerable time. This classic

---

*Senior Lecturer, Graduate School of Business and Public Policy.

organizational structure, although useful in many ways, is also known well to be exceptionally poor at responding to change. In the context of military transformation—the quest for agility and early adaptation, such a problem should be clear and compelling.

14.2 Acquisition Organization: Decision Making

Since the late 1980s, acquisition organizations have operated under a tiered decision-making structure that consists of 1) the program manager (PM, also referred to as the project or product manager), who is responsible for managing a specific program; 2) executives in charge of higher-level programs or organizations; and 3) the milestone decision authority (MDA), who has ultimate decision responsibility for the program. Current policy describes the PM and MDA positions as follows:

> [The PM is] the designated individual with responsibility for and authority to accomplish program objectives for development, production, and sustainment to meet the user's operational needs. The PM shall be accountable for credible cost, schedule, and performance reporting to the MDA.
>
> [The MDA is] the designated individual with overall responsibility for a program. The MDA shall have the authority to approve entry of an acquisition program into the next phase of the acquisition process and shall be accountable for cost, schedule, and performance reporting to higher authority, including Congressional reporting ([2], p. 2).

Managerial levels between the PM and the MDA can consist of 1) a program executive officer (PEO), who is responsible for overseeing a number of related programs (e.g., PEO for ships, PEO for ground combat systems), and 2) an acquisition executive (typically an assistant secretary) of the cognizant military department (Army, Navy, or Air Force).

The specific structure for decision making in any acquisition program will depend on the importance and particular circumstances of that program. The DoD places programs into acquisition categories (ACATs) according to their respective projected expenditures, with ACAT I programs being the most costly (hence, the most important) programs and ACAT III and IV programs being the least costly programs. As one would expect, the most senior officials act as MDAs for ACAT I programs, whereas ACAT III and IV decisions are made by lower-level officials. Figure 14.1 shows the DoD's chain of command for acquisition decision making for the largest ACAT I programs.

The Defense Acquisition Board (DAB), composed of senior DoD and service officials, serves as an advisory body to the Defense Acquisition Executive. The DAB reviews programs, most notably prior to milestone reviews, and makes recommendations on critical program issues.

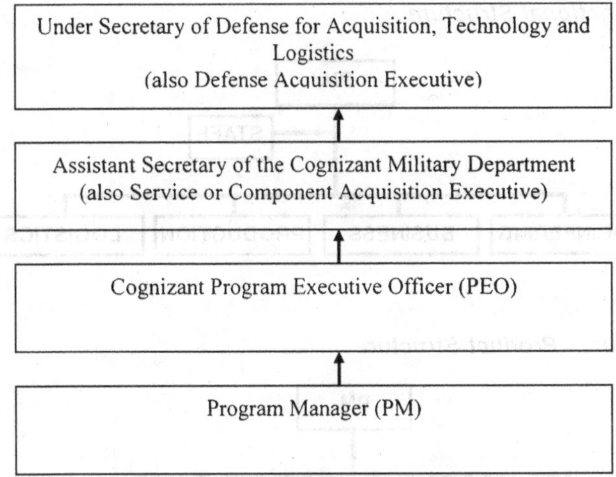

Fig. 14.1 Decision authorities for major defense acquisition programs.

14.3 ACQUISITION ORGANIZATION: PROGRAM MANAGEMENT OFFICES

Although many stakeholders represent different parts of the acquisition enterprise, the Program Management Office (PMO) is the locus of the government's managerial activities. Several alternatives for the organizational structure of PMOs exist. Some offices can be formed with an orientation either toward functional areas or toward products (Fig. 14.2). In most cases, however, PMOs exhibit the matrix organizational form (Fig. 14.3), with both permanently assigned core personnel and temporarily assigned colocated personnel on loan from DoD commodity and systems commands. A significant number of contractor personnel are also often present in the PMO.

Internally, PMOs often organize in ad hoc teams oriented to specific areas of each project (Fig. 14.4). This tendency stems largely from DoD initiatives over the last 10 years to implement integrated product and process development (IPPD) using integrated product teams (IPT). This management philosophy emphasizes the potential of collective knowledge via small organizations with cross-functional or multidisciplinary members [4]. Interestingly, the ideas in this IPPD/IPT philosophy of work implementation and problem-solving are also embodied and magnified in emerging thought regarding command and control in tactical military organizations. Alberts and Hayes' text *Power to the Edge* [5] recognizes the benefit of using information-age technology and low-level empowerment to transfer knowledge and power to the point of an organization's interaction with its environment.

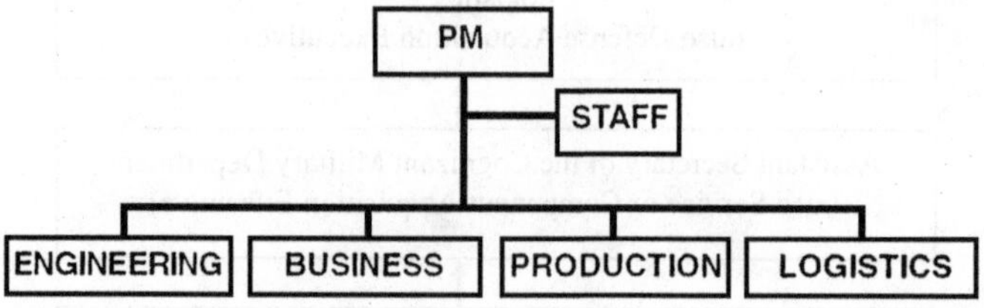

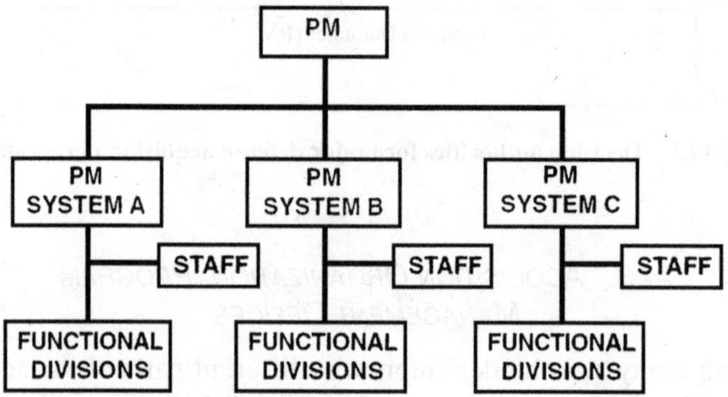

Fig. 14.2 Functional and product-oriented project organizational structures (adapted from [3]).

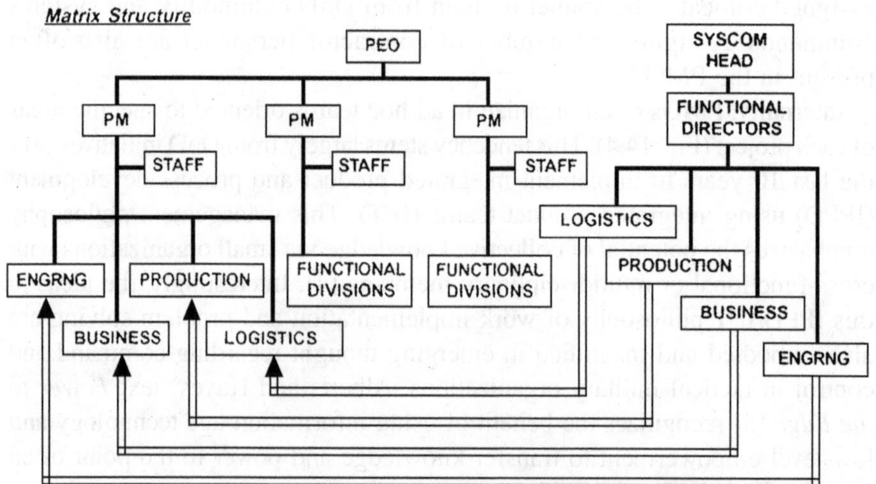

Fig. 14.3 Matrix program management office structure (adapted from [3]).

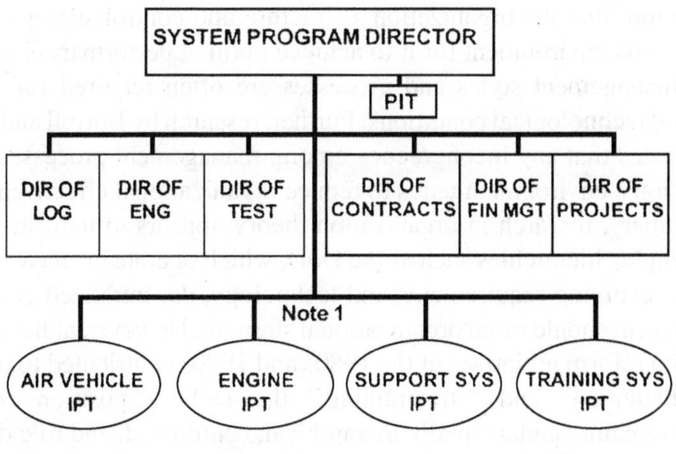

Fig. 14.4 Example (aircraft) PMO IPT structure (adapted from [3]).

14.4 ORGANIZATION THEORY: ACQUISITION APPLICATIONS

In contrast to classical organization theory that posits "one best way" to organize, neoclassical theory holds that organizational structures must change in response to contingencies of size, technology, environment, and other factors. Indeed, it is accepted widely that, when faced with uncertainty (a situation in which procurement personnel have access to less information than is needed), management's appropriate response should be either to redesign the organization for the task at hand or to improve information flows and processing [6]. Van Creveld [7] applies this same principle to command and control of combat elements in war. He argues that the command structure must either create a greater demand for information (vertically, horizontally, or both) and increase the size and complexity of the directing organization, or it must enable the local forces to deal semi-independently with the situation. His central theme is that decentralized control is the superior method of dealing with uncertainty—whether with the task at hand or with transformation of the organization itself. Research by Delbecq et al. [8] has shown further that, as complexity and uncertainty increase, hierarchical management control and vertical communication strategies are considered inferior to the horizontal communication channels incorporated in less formal organizations.

The Law of Requisite Variety [9] states loosely that, in order to cope with the variety of challenges imposed by it, the internal capabilities of a system must be as diverse as those required by its environment. Organizational evolution and survival are dependent upon requisite variety, particularly in environmental contexts that are dynamic and unpredictable. Ashby's Law suggests, too, that the organization's structure and control strategy must be matched to its environment for it to achieve optimal performance. Open and flexible management styles and processes are often required for dynamic market and technological conditions. Further, research by Burrell and Morgan [10] indicates that any incongruence among management processes and the organization's environment tends to reduce organizational effectiveness.

In summary, research in organization theory appears to indicate that, for large, complex hierarchies such as the DoD, which operate in an environment of rapidly evolving requirements and technology, decentralized control and empowerment should be an organizational strength. However, although major acquisition reform initiatives of the 1980s and 1990s contributed to organizational "flattening" and "streamlining," the DoD acquisition command structure remains fundamentally hierarchical, centralized, and rule driven.

14.5 Conclusion: Organizational Control and Project Risk

Although U.S. weaponry is considered some of the best in the world, the major acquisition projects to acquire it are often fraught with cost and schedule growth. Sometimes these weaponry even fail to meet specifications or to provide the capabilities desired. The risks of unfavorable project outcomes run the range from inconsequential to loss of life to, at the extreme, compromises of national security. Clearly, some level of oversight and control of acquisition programs is necessary to reduce the possibility of such adverse outcomes.

Conceptually, investments in project control fall upon a line of diminishing return. Although meetings, reviews, and reports (i.e., more control) will help to measure progress, reveal variance from plans, and determine corrective action, there will be associated costs from these "off-core activities" and diversion of effort from the actual project work to be performed [11]. The general paradigm has been that organizational control measures demand project resources, but serve to buy down risk (Fig. 14.5).

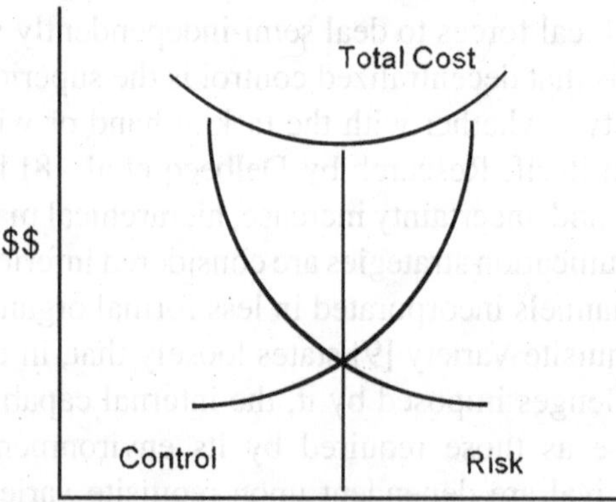

Fig. 14.5 Relationships among project cost, control, and risk.

A better understanding of this important relationship might somewhat benefit senior executives and project managers. The relationship suggests an optimal organizational design that balances control, risk, and cost in each unique endeavor [12].

A one-size-fits-all management policy is probably naïve. If managers can ascertain early on the criticality (and tolerable level) of project quality risk, they can perhaps select along a continuum the level of organizational hierarchy and centralization needed to control project outcomes. Or, in other words, how much will added bureaucracy cost to alleviate risk? The answer will differ—necessarily—for every project.

REFERENCES

[1] Mintzberg, H., *The Structuring of Organizations*, Prentice-Hall, Englewood Cliffs, NJ, 1979.

[2] USD (AT&L), "The Defense Acquisition System," Dept. of Defense Directive 5000.1, Washington, D.C., 12 May, 2003.

[3] Defense Acquisition Univ., *Program Managers' Toolkit*, 14th ed., ver. 2.0, Washington, D.C., Feb. 2005.

[4] Office of the Under Secretary of Defense (Acquisition and Technology), *DoD Integrated Product and Process Development Handbook*, 20301-3000, Washington, D.C., Aug. 1998.

[5] Alberts, D. S., and Hayes, R. E., *Power to the Edge*, Command and Control Research Program (CCRP), Washington, D.C., 2003.

[6] Galbraith, J. R., *Designing Complex Organizations*, Addison Wesley Longman, Reading, MA, 1973.

[7] Van Creveld, M., *Command in War*, Harvard Univ. Press, Boston, MA, 1985.

[8] Delbecq, A. L., Van de Ven, A., and Gustafson, D., *Group Techniques for Program Planning*, 2nd ed., Greenbriar Press, Madison, WI, 1986.

[9] Ashby, W. R., *An Introduction to Cybernetics*, Chapman and Hall, London, 1960.

[10] Morgan, G., *Images of Organization*, Sage Publications, Thousand Oaks, CA, 1997, p. 59.

[11] Wysocki, R. K., *Effective Project Management*, 3rd ed., Wiley, Indianapolis, IN, 2003, p. 54.

[12] Dillard, J. T., and Nissen, M. E., "Determining the Best Loci of Knowledge, Responsibilities and Decision Rights in Major Acquisition Organizations," *Proceedings of the Second Acquisition Research Symposium*, Naval Postgraduate School, Monterey, CA, May 2005, pp. 80–111.

STUDY QUESTIONS

14.1 How is your current organization structured?

14.2 What levels of hierarchy are evident in its decision-making processes?

14.3 To what extent are IPTs used in your organization, and with what degree of success?

14.4 How would you describe the relationship between organizational control and risk in your organization?

Chapter 15

Defense Acquisition Workforce

Rene G. Rendon*
Naval Postgraduate School, Monterey, California

Learning Objectives

- Understand how the Department of Defense defines the acquisition workforce and identify the specific acquisition-related functional areas
- Explain the purpose of the Defense Acquisition Workforce Improvement Act
- Explain the purpose of the Defense Acquisition University
- Understand the purpose of the *AT&L Human Capital Strategic Plan*
- Discuss the attempts by the private sector to professionalize the acquisition workforce
- Explain some of the current challenges related to the defense acquisition workforce

15.1 Introduction

The *Federal Acquisition Regulation* (FAR) states that "participants in the acquisition process should work together as a team and should be empowered to make decisions within their area of responsibility." In addition, the *FAR* states that "the acquisition team consists of all participants in government acquisition, including not only representatives of the technical, supply, and procurement communities, but also the customers they serve and the contractors who provide the products and services" ([1], 1.102). The purpose of this chapter is to provide an overview of the members of this acquisition team, specifically the acquisition workforce within the U.S. Department of Defense (DoD). This chapter will focus on how the Department of Defense manages, organizes, and trains its acquisition workforce to ensure it has the right skills and competencies to successfully manage the DoD's acquisition projects. First,

---

*Senior Lecturer, Graduate School of Business and Public Policy.

a discussion of how the DoD defines the acquisition workforce and identifies the specific acquisition career fields will be presented. Next, the specific legislative actions that have an impact on defense acquisition workforce education, training, and experience requirements will be reviewed. A discussion on how the Department of Defense and industry have attempted to professionalize their acquisition workforces will then be provided, along with some brief highlights of DoD AT&L's human capital strategy. Finally, some current challenges related to the defense acquisition workforce will be identified.

15.2 *Defining the Acquisition Workforce*

Box 15.1 Root Causes of Acquisition Problems

"At a hearing by the Senate Armed Services Committee earlier this month, Senator Carl M. Levin of Michigan said far too many weapons acquisitions had been plagued by 'cost increases, late deliveries to the war fighters and performance shortfalls.' Senator Levin added that 25 of the Pentagon's major defense acquisition programs had overruns of at least 50%. And he expressed concern about an 'alarming lack of acquisition planning across the department.' 'The root cause of these and other problems in the defense acquisition system is our failure to maintain an acquisition work force with the resources and skills needed to manage the department's acquisition system,' Mr. Levin said" [2].

DoD acquisition leaders, along with members of Congress (see Box 15.1), recognize the contributions of a competent workforce to acquisition effectiveness. The DoD Directive 5000.1 states that the Department of Defense "shall maintain a proficient acquisition, technology, and logistics workforce that is flexible and highly skilled across a range of management, technical, and business disciplines." In addition, this directive requires that the USD(AT&L) "shall establish education, training, and experience standards for each acquisition position based on the level of complexity of duties carried out in that position" [3].

One of the main challenges in maintaining a proficient acquisition workforce is actually defining what makes up this workforce. As the DoD increases its outsourcing and contracting out of supplies and services, the nature of defense acquisition is changing to reflect more complex procurements, such as system-of-systems acquisitions, performance-based services acquisition, and consolidated base operations support contracts. As the nature of defense acquisition becomes more complex, the members of the project teams managing these acquisitions also begin to change.

The Department of Defense uses a multidisciplinary, multifunctional definition of its acquisition workforce [4]. The DoD officially refers to its acquisition workforce as the acquisition, technology, and logistics workforce. The DoD AT&L workforce includes contracting, program, technical, budget, financial, logistics, scientific, and engineering personnel. As reflected in Fig. 15.1, seven career fields represent approximately 87% of the total DoD AT&L workforce. These career fields include systems planning, research, development, and engineering (SPRDE); contracting; program management; life-cycle logistics; production, quality, and manufacturing; business, cost estimating, and financial management; and test and evaluation.

In 2006, there were 128,242 members in the defense acquisition workforce. 88% were civilian, and 12% were military ([5], p. 2). Compared to civilian agencies, the DoD workforce is highly educated, experienced, and well trained. Over 72% of DoD civilian acquisition personnel have a bachelor's or higher degree, and 23% have advanced degrees ([5], p. 3). By comparison, less than half of the current federal civilian workforce holds college degrees ([6], p. 3).

15.3 DEFENSE ACQUISITION WORKFORCE IMPROVEMENT ACT

Public Law 101-510 enacted the Defense Acquisition Workforce Improvement Act (DAWIA) in 1990. (The provisions of DAWIA are implemented in the DoD through the DoD Directive 5000.52, the Defense Acquisition, Technology, and Logistics Workforce Education, Training, and Career Development Program, DoD Instruction 5000.55, Reporting Management Information on DoD Military and Civilian Acquisition Personnel and Positions, and DoD Instruction 5000.66, Operation of the Defense Acquisition, Technology and Logistics Workforce Education, Training, and Career Development Program.) The purpose of the act was to improve the effectiveness of the personnel who manage and implement defense acquisition programs. In addition, the act required the establishment of an Acquisition Corps and professionalization of the acquisition workforce through the establishment of education, training, and acquisition-related experience requirements [7]. The DAWIA also required the establishment of a defense acquisition university structure. Under the approval of the USD AT&L, the Defense Acquisition University (DAU) develops curricula for each acquisition career field to include descriptions of the education, experience, and core training required to meet the standards for certification. These education, training, and experience requirements are based on the complexities of the acquisition job. Civilian positions and military billets in the acquisition system have acquisition duties that fall into the career fields as shown in Fig. 15.1.

The DAU training courses are intended to provide students unique acquisition knowledge for specific acquisition workforce assignments, jobs, or positions. In addition, the DAU courses are developed to help the acquisition workforce maintain proficiency and remain current with DoD acquisition legislation, regulation, and policy. Although all defense agencies follow the

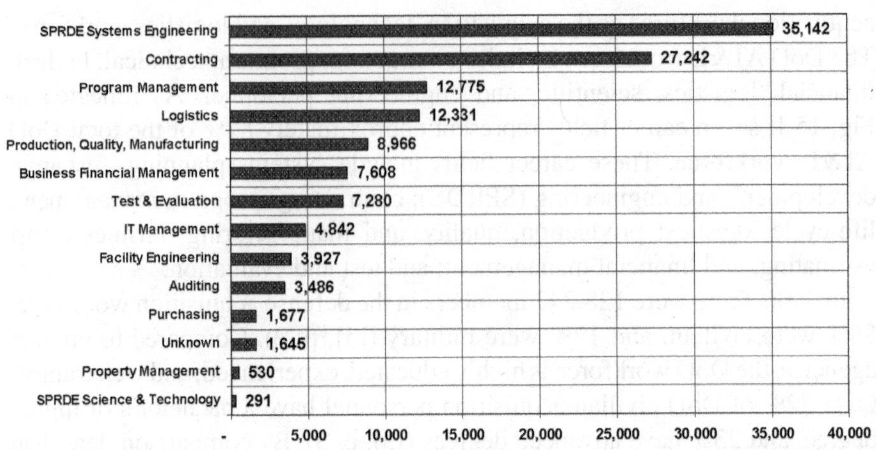

Fig. 15.1 AT&L workforce count by career field (as of September 2006) (adapted from [5]).

DAU curriculum, some civilian agencies (including NASA and the Department of Energy) also follow the DAU curriculum specifically for the contracting and purchasing career fields [4].

Public Law 108-36 enacted changes to the DAWIA in 2003. These changes were focused on providing greater flexibility in managing the AT&L workforce. It also established professional standards for each AT&L career field [5]. A summary of the Defense Acquisition Workforce Improvement Act and its changes is provided in Table 15.1 [8].

15.4 Emphasis on Human Capital

In 2005, the Government Accountability Office (GAO) published its "Framework for Assessing the Acquisition Function at Federal Agencies." The GAO's framework identified four cornerstones that have been shown to promote an efficient, effective, and accountable acquisition function. One of these cornerstones is human capital. In this area, the GAO states that "successfully acquiring goods and services and executing and monitoring contracts to help the agency meet its mission requires valuing and investing in the acquisition workforce. Agencies must think strategically about attracting, developing, and retaining talent, and creating a results oriented culture within the acquisition workforce" ([10], p. viii). The Department of Defense has continued this emphasis on the human capital cornerstone by developing the *AT&L Human Capital Strategic Plan*. The purpose of the *AT&L Human Capital Strategic Plan* is to focus on achieving the Under Secretary of Defense for Acquisition, Technology, and Logistics' goal of a "high performing, agile, and ethical workforce" ([5], p. 2). Figure 15.2 lists the *AT&L Human Capital Strategic Plan* goals.

TABLE 15.1 SUMMARY OF DAWIA HISTORY (ADAPTED FROM [8])

Date	History
November 1990	Congress enacts the Defense Acquisition Worforce Improvement Act (DAWIA, Title 10, U.S.Code, Chapter 87).
October 2000	Fiscal Year (FY) 01 National Defense Authorization Act (NDAA) amends DAWIA, revising the education requirements for 1102s and contracting officers with warrants above the simplified acquisition threshold to require a baccalaureate degree and 24 semester hours in specified disciplines.
December 2002	FY02 NDAA amends DAWIA, expands the 1102 education requirements to members of the armed forces in equivalent occupational specialties and provides for limited expectations to include exceptions for the contingency contracting force and for individuals in developmental positions. The law establishes alternative minimum education requirements for the contingency contracting force and provides authority to establish developmental programs.
November 2003	FY04 NDAA amends DAWIA, providing a number of flexibilities to enable the DoD to more effectively develop and manage the AT&L workforce.
October 2004	FY05 NDAA amends DAWIA, changing Acquisition Corps membership requirements and providing flexibility in the designation of Critical Acquisition Positions (CAPs).
January 2005	OSD issues revised DoDD 5000.52.
December 2005 January 2006	OSD issues DoDI 5000.66 and the DoD Desk Guide for AT&L Workforce Career Management, incorporating statutory changes resulting from FY04 and FY05 NDAA.

Goal 1 — Align and fully integrate with overarching DoD human capital initiatives.

Goal 2 — Maintain a decentralized execution strategy that recognizes the Components' lead role and responsibility for force planning and workforce management.

Goal 3 — Establish a comprehensive, data driven workforce analysis and decision-making capability.

Goal 4 — Provide learing assets at the point of need to support mission-responsive human capital development.

Goal 5 — Execute DoD AT&L Workforce Communication Plan that is owned by all DoD AT&L senior leaders (One Team, One Vision, A Common Message, and Integrated Strategies).

Goal 6 — Recruit, develop, and retain a mission-ready DoD AT&L workforce through comprehensive talent management.

Fig. 15.2 AT&L HCSP version 3.0 goals (adapted from [5]).

The initial version of the *AT&L Human Capital Strategic Plan* was issued in June 2006, and version 3.0 was issued in 2007. It is expected that future revisions to the plan will be published as progress is made toward the preceding goals and as the expected changes in the acquisition workforce continue to materialize.

15.5 EDUCATION AND TRAINING

In addition to the Department of Defense's efforts to improve the effectiveness of its AT&L workforce through the establishment of education, training, and experience requirements, as well as the implementation of the *Human Capital Strategic Plan*, there have been many other initiatives and programs also focused on professionalizing the acquisition workforce. These include the various educational institutions, training organizations, and professional associations.

The DoD's initiatives involving defense acquisition program-management education and training can be traced back to 1964, with the creation of the Defense Weapons System Management Center (DWSMC) at Wright-Patterson Air Force Base in Dayton, Ohio. The mission of the DWSMC was to foster improvement in the quality of program management. The DWSMC was disestablished in June 1971 with the creation of the Defense Systems Management School (DSMS), later to be designated as Defense Systems Management College (DSMC) in Fort Belvoir, Virginia [11]. Today, the Defense Systems Management College School for Program Managers is part of the DAU.

In addition, DoD graduate education schools like the Air Force Institute of Technology (AFIT) and the Naval Postgraduate School (NPS) have been offering graduate education degree programs in acquisition related areas since the 1950s. In 1954, the 83rd Congress authorized the commander, Air University, to confer degrees upon graduates from AFIT. The first undergraduate engineering degrees were granted in 1956, and the first graduate degrees in business in 1958 (data available online at http://www.afit.edu/pa/AFIT_History.cfm). Additionally, a thesis search in the Naval Postgraduate School library catalog finds the earliest acquisition-related graduate thesis published in 1965. Today, the Naval Postgraduate School offers formal defense-focused MBA degree programs with specializations in systems acquisition management, as well as acquisition and contract management. The NPS also offers a defense-focused Master of Science (MS) degree in program management as well as in contract management. These programs are offered in both in-resident and distant-learning formats.

The DAWIA education and training requirements have had a tremendous impact on the number of educational institutions offering formal education programs in support of the AT&L workforce. A quick review of the DAU Web site reveals that a number of academic courses certified to be equivalent to the mandated DAWIA courses are being offered by colleges and universities, government training organizations, as well as private training companies. There are also many civilian training and consulting firms that offer DAWIA-equivalent courses to the DoD AT&L workforce, as well as the contractor community. The complete list of colleges and universities offering training courses that have been granted DAWIA equivalencies is found at http://www.dau.mil/learning/appg.aspx.

15.6 PROFESSIONAL ASSOCIATIONS

Professional associations have also played a significant part in professionalizing the defense acquisition workforce. These associations provide professional development opportunities, including educational conferences, workshops, and certification programs. Although there are a variety of professional associations that represent the various acquisition workforce career fields listed on the AT&L Knowledge Sharing System (data available online at https://akss.dau.mil/Lists/Education%20Training/Professional%20Organizations.aspx), some of the predominant ones will be discussed next.

The program-management career field is represented by the Project Management Institute (PMI). The PMI is the leading membership association for the project-management profession. With more than 260,000 members in over 170 countries, the PMI is actively engaged in advocacy for the profession, setting professional standards, conducting research, and providing access to a wealth of information and resources. The PMI also promotes career and professional development and offers certification, networking, and community-involvement opportunities. The PMI offers professional credentials, including the Project Management Professional (PMP®) and the Program Management Professional (PgMP℠) (data available online at www.pmi.org).

The contracting career field is well represented by the National Contract Management Association (NCMA). The NCMA exists to enable the workforce to grow professionally, to assess individual and organizational competency against professional standards, establish values, develop best practices, and to provide access to skilled individuals—enabling enterprises to improve their buyer–seller relationships. The NCMA's vision is to lead and represent the contract-management profession through improved buyer–seller relationships based on common values, practices, and professional standards. The NCMA offers various professional certifications, including the Certified Professional Contracts Manager (CPCM), Certified Commercial Contracts Manager (CCCM), and the Certified Federal Contracts Manager (CFCM) (data available online at www.ncmahq).

The supply-management profession is represented by the Institute for Supply Management. Founded in 1915, the Institute for Supply Management™ (ISM) is the largest supply-management association in the world as well as one of the most respected. ISM's mission is to lead the supply-management profession through its standards of excellence, research, promotional activities, and education. ISM's membership base includes more than 40,000 supply-management professionals with a network of domestic and international affiliated associations. ISM is a not-for-profit association that provides opportunities for the promotion of the profession and the expansion of professional skills and knowledge. ISM offers professional credentials including the Certified Professional in Supply Management (CPSM) and the Certified Purchasing Manager (C.P.M.) (data available online at http://www.ism.ws).

The financial-management career field is represented by the American Society of Military Comptrollers (ASMC). The ASMC is the nonprofit educational and professional organization for persons (military and civilian) involved in the overall field of military comptrollership. The ASMC promotes the education and training of its members and supports the development and advancement of the profession of military comptrollership. The society sponsors research, provides professional programs to keep members abreast of current issues, and encourages the exchange of techniques and approaches. The ASMC awards the Certified Defense Financial Manager (CDFM) certification to those individuals who meet the experience and education requirements and pass the required examinations (data available online at http://www.asmconline.org).

> The cost-management profession is represented by the Association for the Advancement of Cost Engineering, International. Since 1956, AACE International has been the leading-edge professional society for cost estimators, cost engineers, schedulers, project managers, and project control specialists. With more than 5,500 members worldwide, AACE International is the largest organization serving the entire spectrum of cost-management professionals. AACE International offers various professional certifications including the Certified Cost Consultants (CCC)/Certified Cost Engineers (CCE), Interim Cost Consultants (ICC), Planning & Scheduling Professionals (PSP), and Earned Value Professionals (EVP) (data available online at http://www.aacei.org/).

The logistics-management community is represented by SOLE—The International Society of Logistics (SOLE). SOLE was founded in 1966 as the Society of Logistics Engineers "to engage in educational, scientific, and literary endeavors to advance the art of logistics technology and management." SOLE is a nonprofit international professional society composed of individuals organized to enhance the art and science of logistics technology, education, and management. The society's professional certification and recognition programs recognize the professional stature and accomplishments of logisticians within commerce, industry, defense, international, federal, and local government agencies, as well as in academic and private institutions. SOLE offers the Certified Master Logistician (CML) and the Certified

Professional Logistician (CPL) certifications (data available online at http://www.sole.org/).

In addition to professional associations representing specific acquisition career fields, the acquisition workforce has also benefited from professional associations representing the various industries involved in defense acquisitions, such as the Aerospace Industries Association of America, Inc. (AIA), the American Institute of Aeronautics and Astronautics (AIAA), and the National Defense Industries Association (NDIA), to name a few.

The Aerospace Industries Association (AIA), founded in 1919, is the premier trade association representing the nation's major aerospace and defense manufacturers. AIA is led by a board of governors consisting of senior representatives of member companies, and an executive committee. A hallmark of AIA is that it receives its policy guidance from the direct involvement of CEO-level officers of the country's major aerospace companies. The government frequently seeks advice from AIA on issues, and AIA provides a forum for government and industry representatives to exchange views and resolve problems on noncompetitive matters related to the aerospace industry (data available online at http://www.aia-aerospace.org).

The American Institute of Aeronautics and Astronautics (AIAA) was formed in 1963 by the merger of two societies, the American Rocket Society, which had begun in 1930 as the American Interplanetary Society, and the Institute of the Aerospace Sciences, established in 1932 as the Institute of the Aeronautical Sciences. The mission of AIAA is to advance the arts, sciences, and technology of aeronautics and astronautics, and to promote the professionalism of those engaged in these pursuits. AIAA encourages original research, furthers dissemination of new knowledge, fosters the professional development of those engaged in science and engineering, improves public understanding of aerospace and its contributions, fosters education in engineering and science, promotes communication among engineers, scientists, and other professional groups, and stimulates outstanding professional accomplishments (data available online at http://www.aiaa.org).

The National Defense Industries Association (NDIA) traces its history to the American Defense Preparedness Association (ADPA), founded in 1919, and the National Security Industrial Association (NSIA), founded in 1944. NDIA was founded in March 1997 as a nonpartisan, nonprofit, educational association. NIDIA's mission is to advocate cutting-edge technology and superior weapons, equipment, training, and support for the war-fighter and first responder, promote a vigorous, responsive, government–industry national security team, and provide a legal and ethical forum for exchange of information between industry and government on national security issues (data available online at http://www.ndia.org).

A cursory review of the DAWIA career fields, as well as the professional associations listed in this chapter, exposes an interesting finding: the DAWIA career fields and professional associations represent every aspect of the defense acquisition environment—except one. As discussed in Chapter 1, there are three major decision-making support systems that impact the management of defense acquisition projects. These are requirements (Do we have a valid requirement?), acquisition strategy (Do we have a strategy for acquiring the requirement?), and resources (Do we have adequate resources to execute the strategy for acquiring the requirement?). The requirements management area is one of the most troublesome aspects of defense-acquisition projects. Recent GAO reports have identified requirements management, specifically requirements and funding instability, as the biggest obstacles to project success ([12], p. 61). The requirements management process is represented by the community of end users and operators of the major weapon systems acquired by the DoD. These end users and operators are responsible for determining the weapon-system requirements for executing the DoD mission. Although requirements management is critical in the DoD acquisition process, it is noticeably absent from the list of DAWIA acquisition career fields; it is also not represented in the professional associations that support the acquisition workforce. It is this author's opinion that requirements management will continue to be an obstacle to DoD project success until the DoD requirements community is included in the DAWIA education, training, and experience standards, or unless it develops its own standards of professionalism.

The professionalization of the acquisition workforce will most likely continue to parallel the transformation of the defense acquisition system. But the jury is still out on the effectiveness of this professionalization process or on the future direction of defense acquisition workforce professionalization initiatives. It has been suggested that the result of professionalization of the acquisition workforce in general, and of DAWIA specifically, is leading to an acquisition workforce that is expert and specialized, yet insular and careerist [13]. Other suggestions on improving the acquisition workforce, specifically the recruitment process, include implementing a Reserve Officer Training Corps (ROTC) program that would train cadet officers in defense acquisition in addition to the regular college curriculum [14]. An even more radical approach to recruiting a professionalized acquisition workforce includes creating a business-management development program within the military academies that would train acquisition cadets and graduate acquisition professionals [14]. No one knows what lies ahead in terms of professionalizing the acquisition workforce. But it remains clear that as defense acquisition management continues to encounter problems in meeting cost, schedule, and performance objectives, the defense acquisition system will continue to be reformed, and the defense acquisition workforce will need to continue to reflect the changing knowledge, skills, and abilities needed to manage defense acquisition programs.

15.7 CONCLUSION: CURRENT CHALLENGES IN THE DEFENSE ACQUISITION WORKFORCE

In 1990, the DAWIA was enacted. That same year, the GAO (now the Government Accountability Office) designated DoD weapon systems acquisition as a "high-risk" area. (DoD contract management was placed on the GAO High Risk List only two years later [12].) As this textbook goes to print, the defense acquisition workforce will be celebrating 18 years of high-risk status for defense weapon-systems acquisition under DAWIA governance.

The future of the DoD acquisition management has many challenges and opportunities. Major defense weapon systems continue to experience cost overruns, schedule delays, and performance problems. The number of contractor protests has increased within the past number of years. Additionally, the number of defense contracts (for both services and systems) continues to increase, without a corresponding increase in the defense acquisition workforce. Box 15.2 illustrates this challenge. The DoD will find it difficult to manage the future AT&L workforce. The current challenges to planning, shaping, and managing the defense acquisition workforce as identified by the DoD include the following: 1) potential loss of retirement-eligible personnel and their knowledge, 2) understanding the differences in workforce generations, and 3) the depleting U.S. workforce pool with increasing competition for talent.

The first challenge ("potential loss of retirement-eligible personnel and their knowledge") will require the DoD to focus on both recruiting and

BOX 15.2 DISTRESSED AND FRUSTRATED WORKFORCE

"'The Defense Department's civilian acquisition workforce has shrunk by about 40% since the early 1990s and now has about 270,000 employees, according to Pentagon statistics and Government Accountability Office reports. Yet defense spending on service contracts increased 78%, to $151 billion, from 1996 to 2006, the reports said. There are 7.5 million federal contractors, 1.5 million more than in 2002, without a corresponding increase in government officials to oversee them,' said Paul C. Light, a public service professor at New York University.

'The acquisition workforce couldn't be in any more distress right now, and I know they are frustrated that they can't oversee the contracts that they have,' Light said. 'They are looking at the hunks of money flowing out but don't have the bodies to keep up. The shortfall, which the government has relied largely on private contractors to fill, has contributed to cost overruns and delays, according to government reports and audits [9].'"

retaining acquisition workforce members and on maintaining organizational competency and process capability in acquisition management. Box 15.3 illustrates this challenge. As the retirement-eligible personnel (and their knowledge) retire and walk out the doors of AT&L acquisition organizations, most of the tacit knowledge needed for managing defense acquisition projects will also exit the organization with them. To maintain organizational competency and process capability in acquisition management, the AT&L acquisition organizations must try to transfer and share the knowledge of the experienced, yet retiring personnel, to the more junior personnel. More importantly, the organizations must also try to implant this knowledge into the organization's acquisition management processes.

BOX 15.3 FEDERAL AGENCY RECRUITMENT

"Over the next two years, federal agencies expect to hire nearly 193,000 new workers in almost every occupational field, according to a report from the Partnership for Public Service, a nonprofit group that urges young Americans to consider careers in government. More than 83,000 of the jobs are expected to be filled at Defense and Homeland Security as part of the continuing effort to secure the nation from terrorism" [15].

The DoD's success in meeting the second and third challenges ("understanding the differences in workforce generations" and "the depleting U.S. workforce pool with increasing competition for talent") will determine if the Department can successfully respond to the first challenge. Figure 15.3 provides an excellent illustration of the generational differences in the U.S. workforce. As the DoD tries to entice the Generation X and Generation Y workforce to join AT&L acquisition organizations, the Department will need to consider and cater to the preferred work environment and motivations of these generations.

	Silent	Baby Boomer	Generation X	Generation Y
Preferred Work Environment	• Promotions comes with seniority • Younger workers should pay their dues • Value sacrifice, conformity, and patience	• Love to have meetings • Job security • Learn technology • Position = respect • Younger workers should pay their dues • Value "face time"	• Fun environment • Use technology • Internal mobility • Flexible schedules • Peers does not equal family • Changing challenges and responsibilities	• Fun environment • Assume technology • Internal mobility • Flexible schedules • Peers = family • Expect bosses to assist and mentor
Motivated by ...	• Satisfaction is a job well done • Being respected	• Advancement • Title recognition • Being valued and needed • Money	• Freedom • Removal of rules • Continuous learning • Time off • Money	• Meaningful work • Working with bright people • Increased responsibility • Time off • Money

Fig. 15.3 Generational differences in the U.S. workforce (adapted from [5]).

REFERENCES

[1] *Federal Acquisition Regulation (FAR)*, Washington, D.C., www.arnet.gov/far [retrieved Nov. 2007].

[2] Schmitt, E., and Thompson, G., "Top Air Force Official Dies in Apparent Suicide," *New York Times*, 15 Oct. 2007, http://www.nytimes.com/2007/10/15/us/15cnd-contract.html?ex=1350100800&en=1b88bfba152069e&ei=5088&partner=rssnyt&emc=rss [retrieved 16 Oct. 2007].

[3] Dept. of Defense (DoD) USD (AT&L), "The Defense Acquisition System," Dept. of Defense Directive 5000.1, Washington, D.C., May 2003.

[4] Government Accountability Office, "Acquisition Workforce: Agencies Need to Better Define and Track the Training of Their Employees," GAO 02-737, Washington, D.C., July 2002.

[5] Dept. of Defense (DoD) USD (AT&L), *AT&L Human Capital Strategic Plan*, ver. 3.0, Washington, D.C., 2007.

[6] Congressional Budget Office, "Characteristics and Pay of Federal Civilian Employees," Washington, D.C., March 2007, p. 3.

[7] Defense Acquisition Workforce Improvement Act, Title 10, U.S.Code, Chap. 87.

[8] Dept. of Defense (DoD) USD (AT&L), "A Desk Guide for Acquisition, Technology, and Logistics Workforce," Washington, D.C., 10 Jan 2006.

[9] Merle, R., "Government Short of Contracting Officers, Officials Struggle to Keep Pace with Rapidly Increasing Defense Spending," *Washington Post*, Thursday, 5 July 2007, p. E08.

[10] Government Accountability Office, "Framework for Assessing the Acquisition Function at Federal Agencies," GAO-05-218G, Washington, D.C., Sept. 2005.

[11] Acker, D., *A History of the Defense Systems Management College*, Defense Systems Management College, Fort Belvoir, VA, 1986, pp. 3–29.

[12] Government Accountability Office, "High-Risk Series: An Update," GAO-07-310, Washington, D.C., Jan. 2007.

[13] Snider, K. F., "DAWIA and the Price of Professionalism," *Acquisition Review Quarterly*, Defense Acquisition Univ., Fort Belvoir, VA, 1996.

[14] Gill, J. H., "Crisis in the Acquisition Workforce," *Acquisition Review Quarterly*, Defense Acquisition Univ., Fort Belvoir, VA, 2001.

[15] Barr, S., "Looking for Quite a Few Good Men and Women," *Washington Post* [Online version], 10 July 2007, p. D04, washingtonpost.com [retrieved 10 July 2007].

STUDY QUESTIONS

15.1 Describe how the Department of Defense defines the acquisition workforce and identify the specific acquisition-related functional areas.

15.2 Explain the purpose of the Defense Acquisition Workforce Improvement Act.

15.3 What is the role of the Defense Acquisition University?

15.4 Discuss the purpose of the *AT&L Human Capital Strategic Plan*.

15.5 Discuss the attempts by the DoD and the private sector to professionalize the acquisition workforce.

15.6 Give examples of professional associations and private-sector associations that support the professionalization of the acquisition workforce.

15.7 Discuss some of the current challenges related to the defense acquisition workforce.

INDEX

Supporting Materials

Many of the topics introduced in this book are discussed in more detail in other AIAA publications. For a complete listing of titles in the Library of Flight, as well as other AIAA publications, please visit http://www.aiaa.org.